Using Australian Money

Book A

All rights reserved. No part of this book may be reproduced or transmitted in any form or by any means without written permission of the author.

Author: Christine R. Draper

© Warru Press, 2023

Published: Warru Press, Rockingham Western Australia, 2023

ISBN: 978-1-922819-04-8

ISBN: 978-1-922819-04-8 © Warru Press, 2023

Recognising Australian Coins

Circle the coin that matches the amount.

Recognising Australian Coins

Colour (or circle) the $1 and $2 coins gold. Colour (or circle) the other coins silver.

How many?

- Gold coins _____
- Silver coins _____

Recognising Australian Coins

Colour or circle the coins and then count them:

- Colour the 5 cent coins blue   There are _____ 5c coins.
- Colour the 10 cents coins green   There are _____ 10c coins.
- Colour the 20 cent coins red   There are _____ 20c coins.
- Colour the 50 cent coins purple   There are _____ 50c coins.
- Colour the 1 dollar coins yellow   There are _____ $1 coins.
- Colour the 2 dollar coins orange   There are _____ $2 coins.

Australian Notes

Colour the Australian notes:

From time to time the images on Australian notes and coins change.

Fifty cent coins are often made to commemorate different occasions. Here are some different fifty cent coins. Notice the first one is round.

  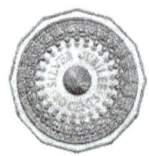    

Ordering Australian Money

Cut out the coins from page 60 or from https://warrupress.com/using-australian-money/ and glue them from smallest value to largest.

Cut out the notes from page 60 and glue them from smallest value to largest.

Put all the money from smallest value to largest.

The smallest and largest coin

Circle the coin or note of the smallest value shown in blue and the largest in orange.

Ordering Australian Money

Cut out the coins from page 61 or from https://warrupress.com/using-australian-money/ and glue them from smallest value to largest.

Cut out the notes from page 61 and glue them from smallest value to largest.

Put all the money from smallest value to largest.

Adding $1 coins

Add each row of $1 coins.

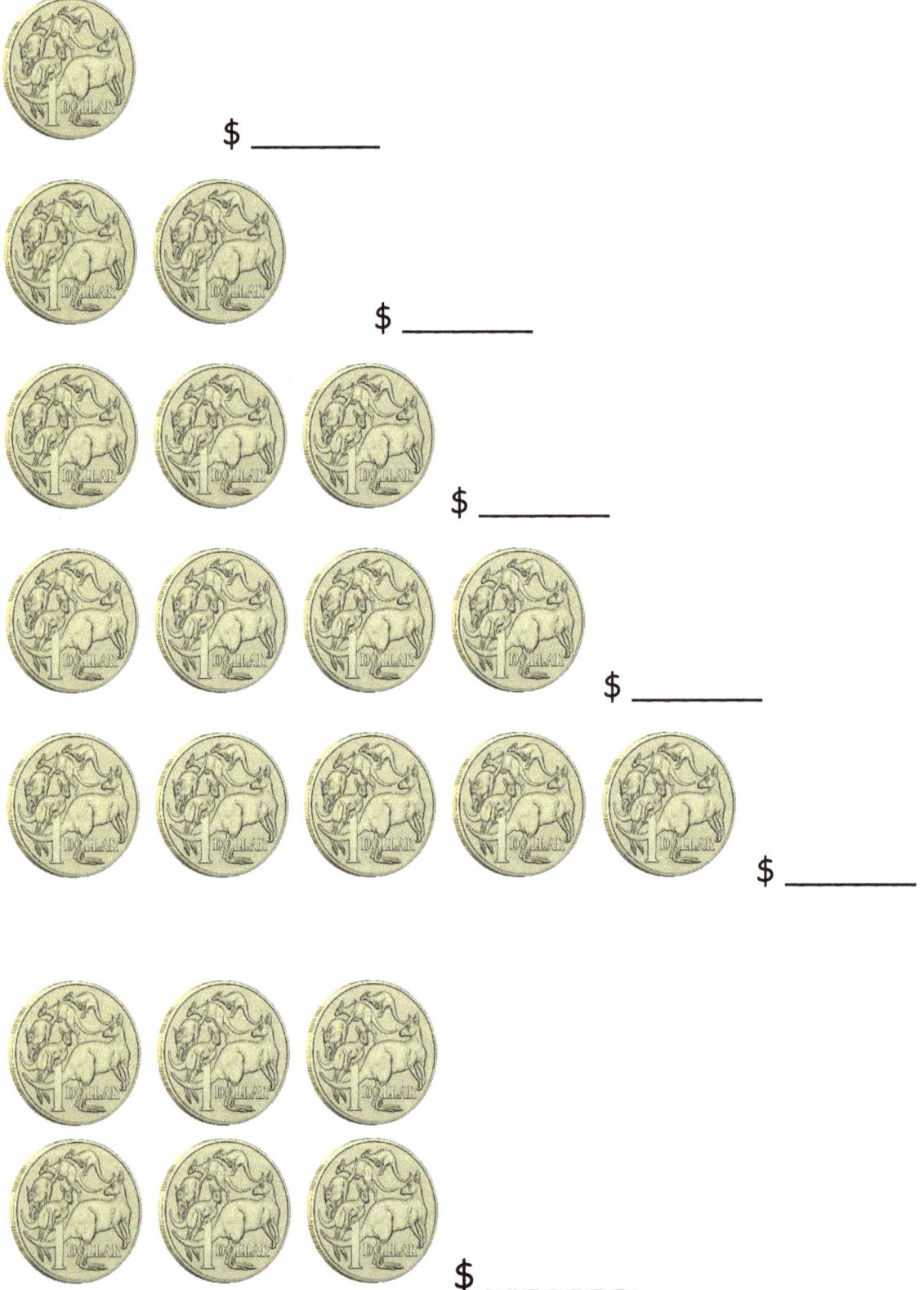

Adding $1 coins

Add each row of $1 coins.

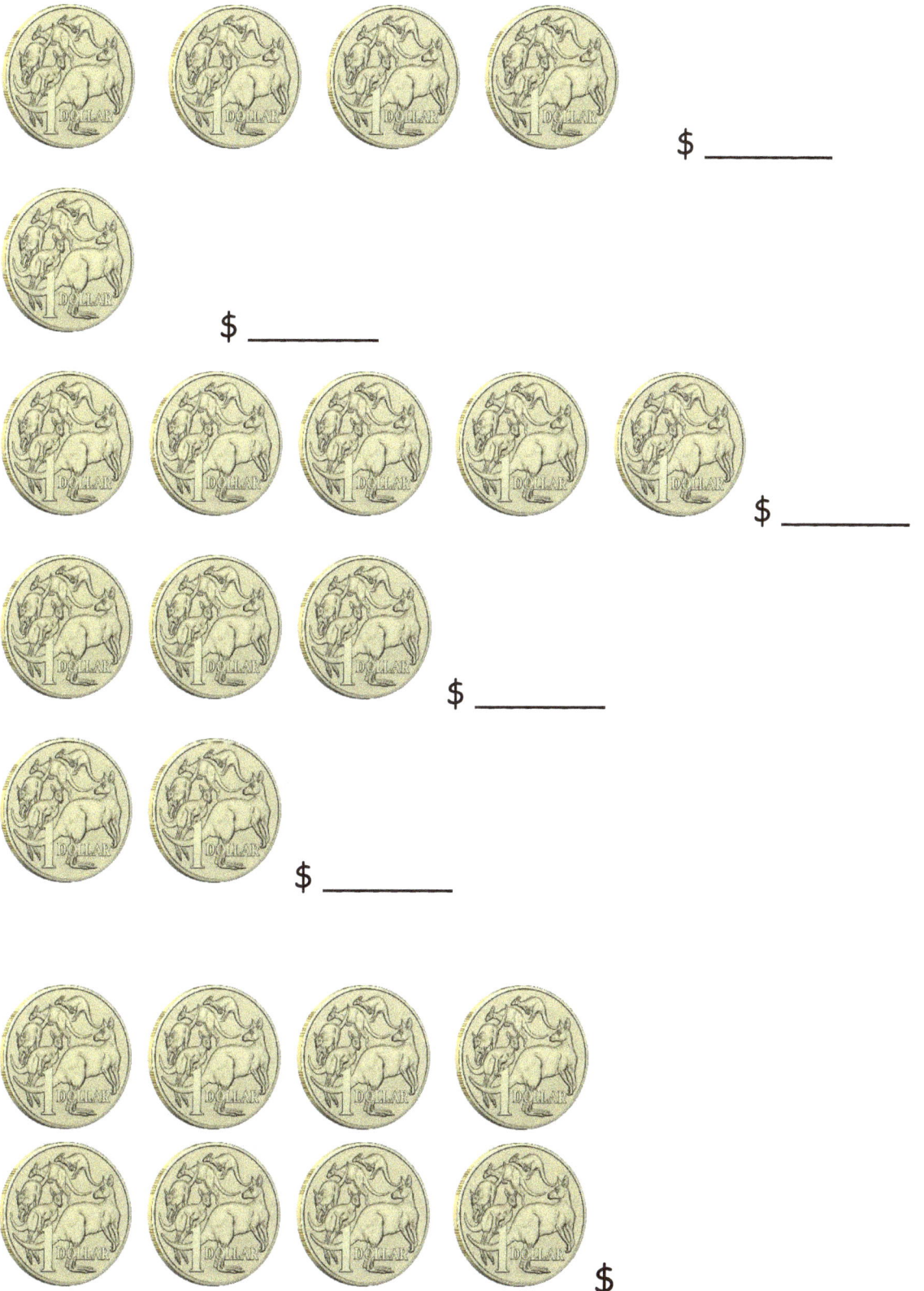

$ _____

$ _____

$ _____

$ _____

$ _____

$ _____

Adding $2 coins

Add each row of $2 coins.

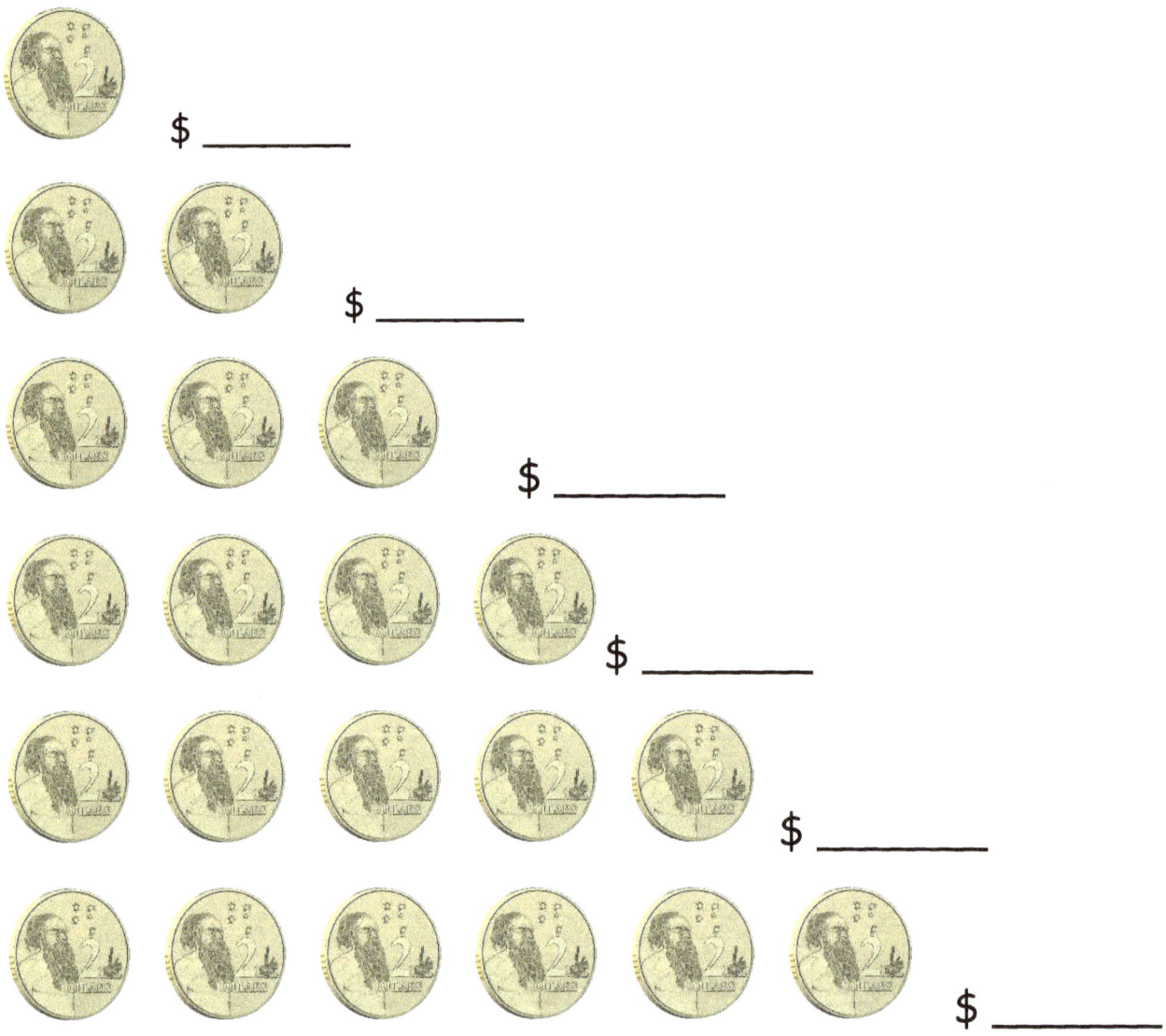

Adding $2 coins

Add each row of $2 coins.

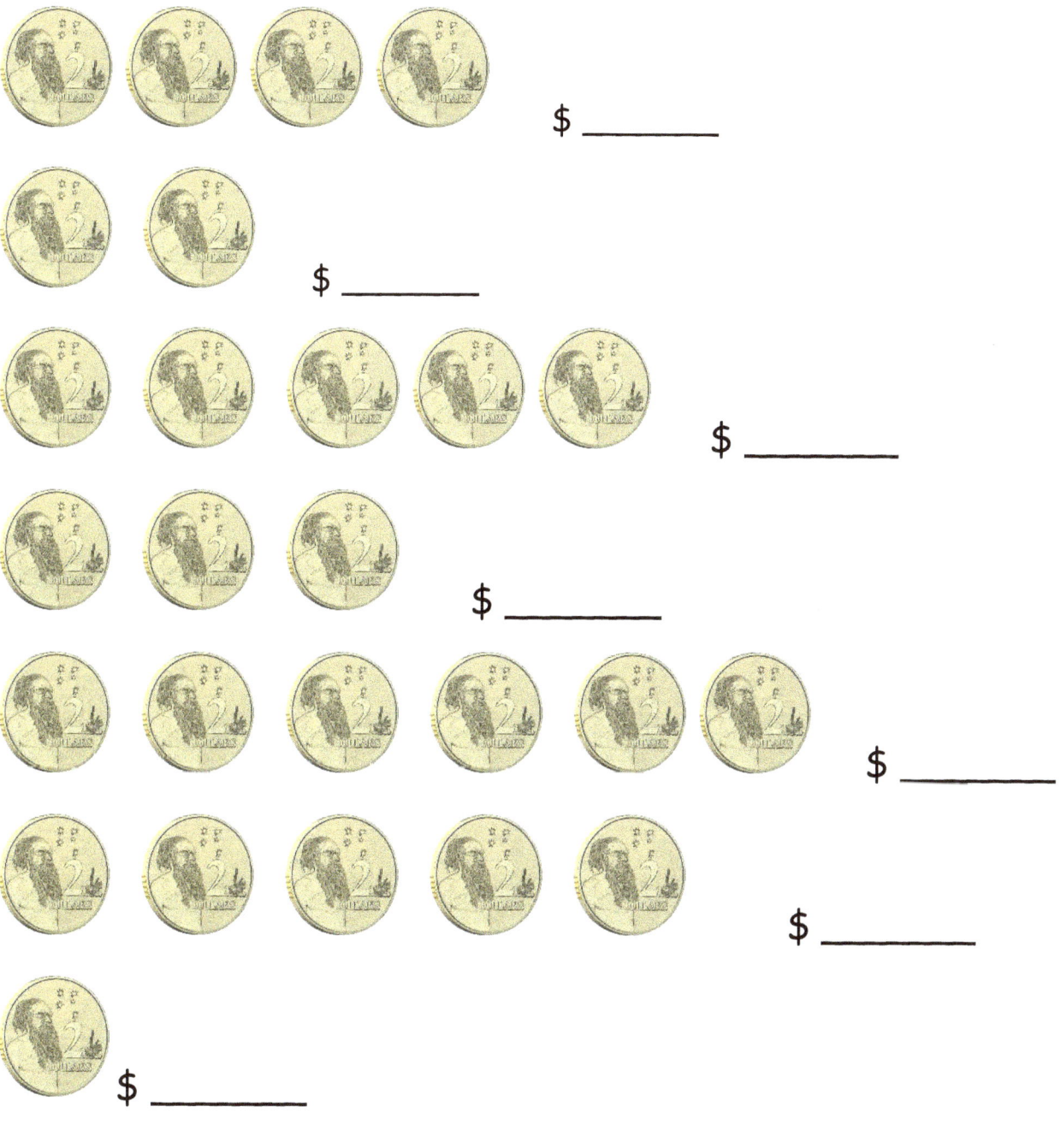

$ _____

$ _____

$ _____

$ _____

$ _____

$ _____

$ _____

Adding $1 and $2 coins

Add each row of coins.

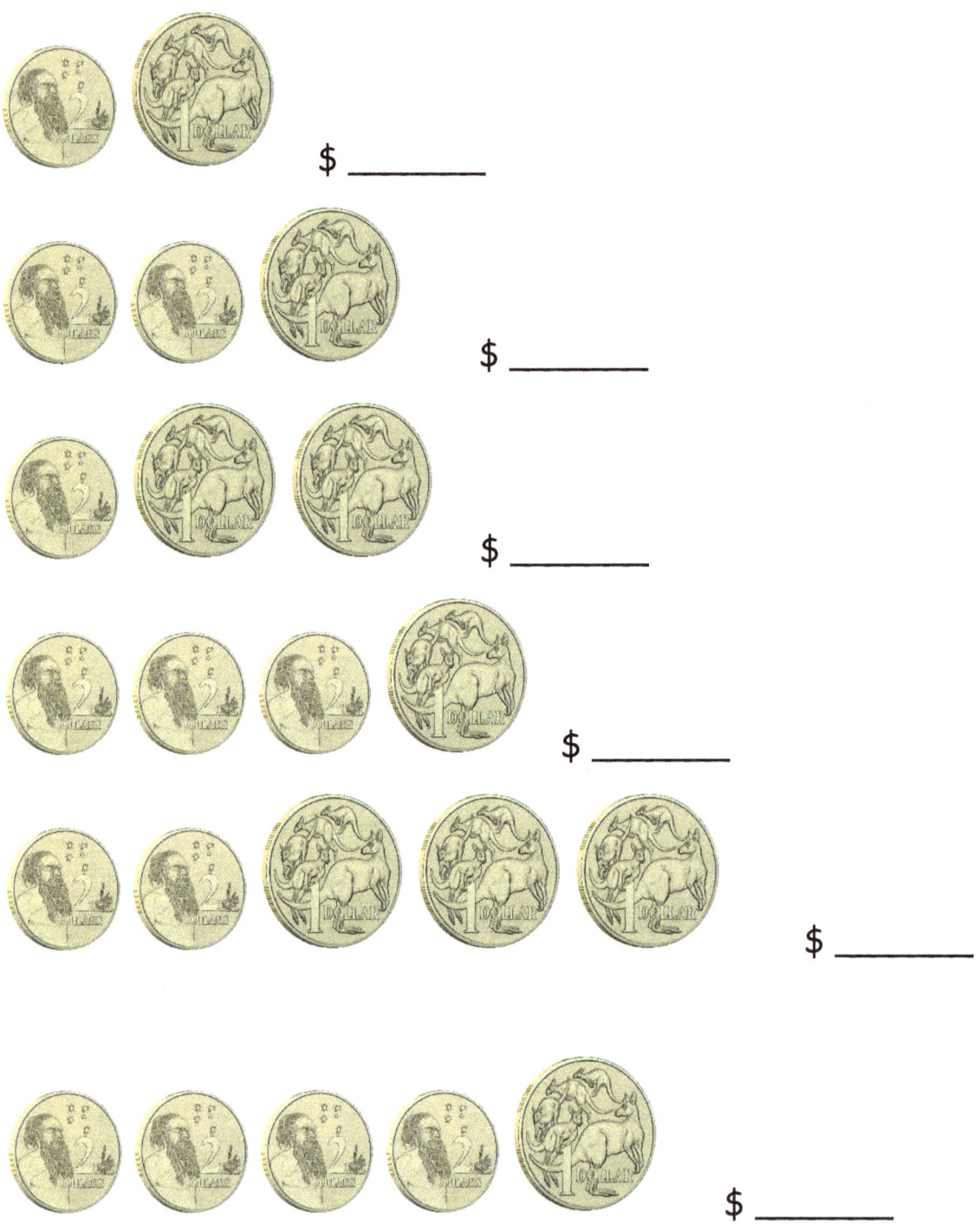

$ _____

$ _____

$ _____

$ _____

$ _____

$ _____

Dollars and cents

100 cents is the same as one dollar.

100 c = $1

For each of the amounts below, circle the dollars and underline the cents.

Hint: Some don't have both dollars and cents shown.

Dollars and cents

100 cents is the same as one dollar.

100 c = $1

How many dollars in the following amounts? Write the amounts as dollars. The first one is done for you.

300c = $3						100c = ____

500c = ____					800c = ____

900c = ____					200c = ____

400c = ____					600c = ____

700c = ____					300c = ____

1000c = ____					2000c = ____

1100c = ____					1700c = ____

1900c = ____					2100c = ____

1500c = ____					3000c = ____

3700c = ____					10000c = ____

Place Value

Every place value is worth ten times more than the place on the right. Here are some of the place values that you need to know.

Tens	Units
1	2

The number above is 12 or twelve. It has one in the tens' place, two in the units' place.

Answer the following questions. The first one is done for you.

15c What digit is in the units place? 5

$36 What digit is in the tens place? _____

$21 What digit is in the units place? _____

89c What digit is in the tens place? _____

54c What digit is in the units place? _____

$54 What digit is in the tens place? _____

87c What digit is in the units place? _____

$10 What digit is in the units place? _____

$10 What digit is in the tens place? _____

96c What digit is in the tens place? _____

69c What digit is in the units place? _____

Adding $10

If we add ten, we are adding one digit to the tens place.

So, if we add 10 dollars to an amount of money in dollars, we add one to the tens place and the units place stays the same.

Example:

Add $10 to $25

Tens	Units
2	5

There is a 2 in the tens place. We add one to the two.

There is a 5 in the units place. That stays the same.

$25 + $10 = $35

Example:

Add $10 to $5

Tens	Units
0	5

There is nothing in the tens place. We add one to zero.

There is a 5 in the units place. That stays the same.

$5 + $10 = $15

Complete these sums by adding $10.

$7 + $10 = _____

$19 + $10 = _____

$36 + $10 = _____

$48 + $10 = _____

$72 + $10 = _____

Adding 10 cents

If we add ten, we are adding one digit to the tens place.

So, if we add 10 cents to an amount of money in cents, we add one to the tens place and the units place stays the same.

Complete the sums by adding ten cents. The first one has been done for you.

36c + 10c = 46c

81c + 10c = _____

29c + 10c = _____

71c + 10c = _____

3c + 10c = _____

58c + 10c = _____

39c + 10c = _____

77c + 10c = _____

16c + 10c = _____

41c + 10c = _____

63c + 10c = _____

Decimal Point

After the units place is a decimal point. If there are no digits after the decimal place it might not be shown. If a decimal point is used in money, we always have two digits after the decimal point.

That means five dollars can be written as $5.00 or $5. It is the same thing.

Write the following amounts using a decimal point. Remember the dollar sign. The first one has been done for you.

$23 = $23.00

$1 = _____

$7 = _____

$8 = _____

$19 = _____

$47 = _____

$56 = _____

$62 = _____

Write the following amounts without a decimal point.

$19.00 = $19

$3.00 = _____

$6.00 = _____

$12.00 = _____

$27.00 = _____

$38.00 = _____

Place Value - hundredths

In money we need to know the first two places after the decimal point. They are the tenths and the hundredths.

Every place value is worth ten times more than the place on the right. Here are some of the place values that you need to know.

Tens	Units	tenths	hundredths
4	3	2	1

The number above is 43.21 or forty-three point two one. If this was in dollars it would be $43.21 or forty three dollars and twenty one cents.

All the squares below contain 100 small squares. So, each part is $\frac{1}{100}$, or 0.01 (one in the hundredths place). This is the same as having 100 cents in a dollar and every cent being $\frac{1}{100}$ of a dollar or $0.01.

For all the squares below write the fraction and the decimal.

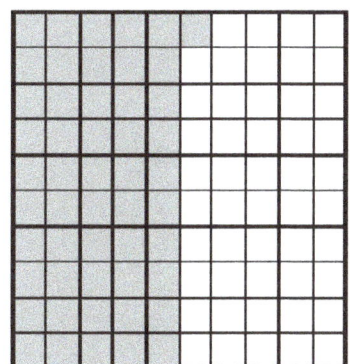

Squares shaded: 10
Fraction: $\frac{10}{100}$
Decimal: 0.10

Squares shaded:
Fraction:
Decimal:

Squares shaded:
Fraction:
Decimal:

Place Value – hundredths

For all the squares below write the fraction and the decimal.

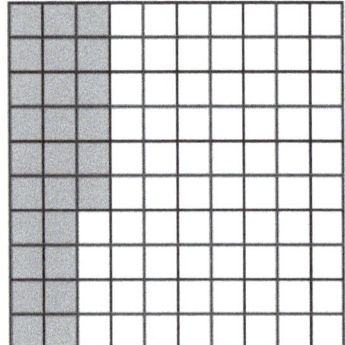

Squares shaded:
Fraction:
Decimal:

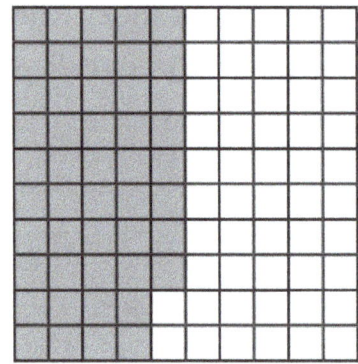

Squares shaded:
Fraction:
Decimal:

Squares shaded:
Fraction:
Decimal:

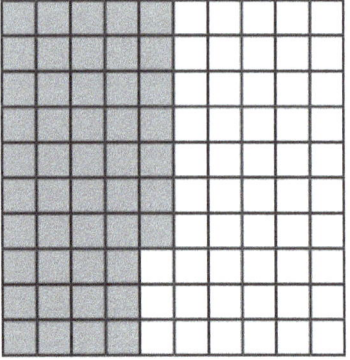

Squares shaded:
Fraction:
Decimal:

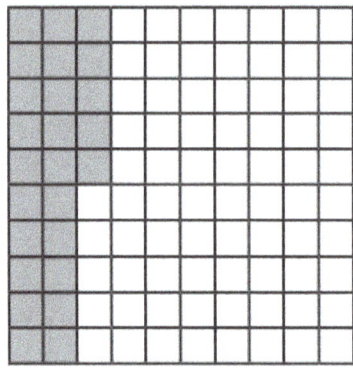

Squares shaded:
Fraction:
Decimal:

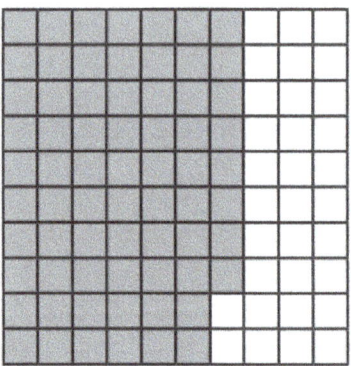

Squares shaded:
Fraction:
Decimal:

Writing cents as dollars

$$1c = \$\frac{1}{100} = \$0.01$$

Write the following amounts as dollars. The first one has been done for you.

68c = $0.68

72c = _____

99c = _____

42c = _____

27c = _____

13c = _____

30c = _____

5c = _____

Write the following amounts as cents. The first one has been done for you.

$0.11 = 11c

$0.68 = _____

$0.15 = _____

$0.50 = _____

$0.24 = _____

$0.82 = _____

$0.22 = _____

$0.75 = _____

$0.02 = _____

Adding 50c coins

Two fifty cent coins equals one dollar.

One fifty cent coin can be written as 50c or $0.50.

Circle groups of two coins then work out what each row is worth. The first one is done for you.

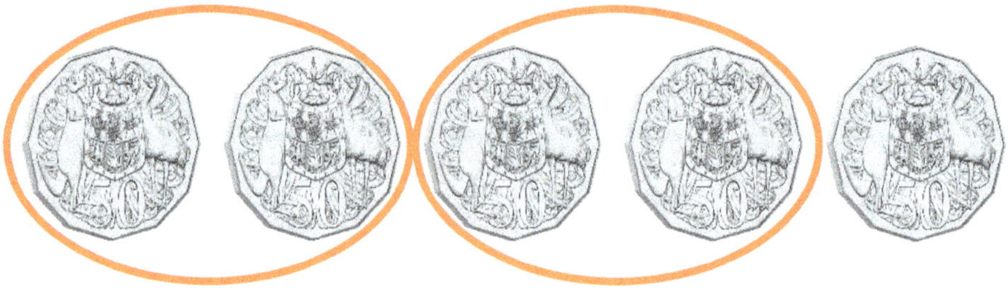

$ 2.50

$ _____

$ _____

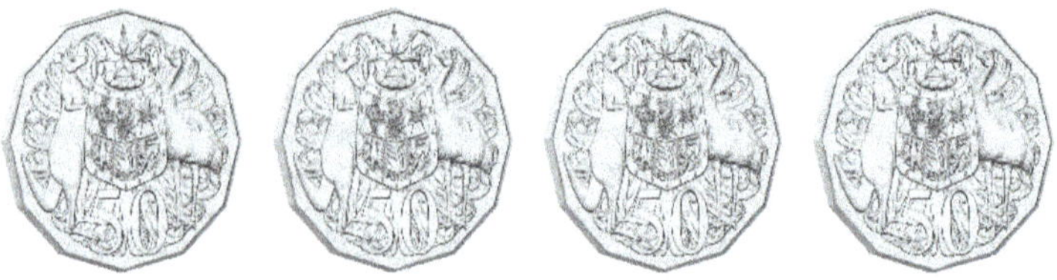

$ _____

$ _____

Adding 50c coins

Two fifty cent coins equals one dollar.

One fifty cent coin can be written as 50c or $0.50.

Circle groups of two coins then work out what each row is worth.

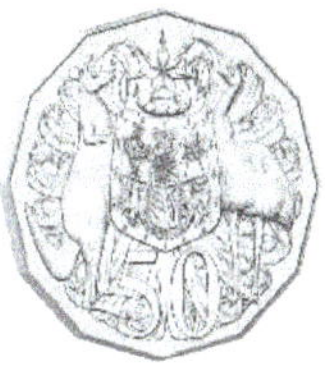

$ _____

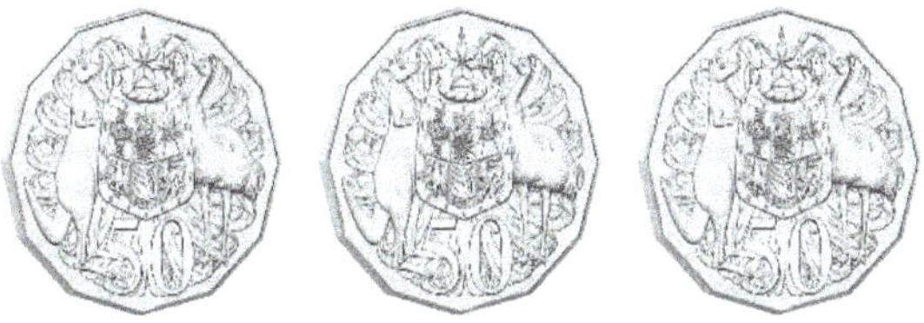

$ _____

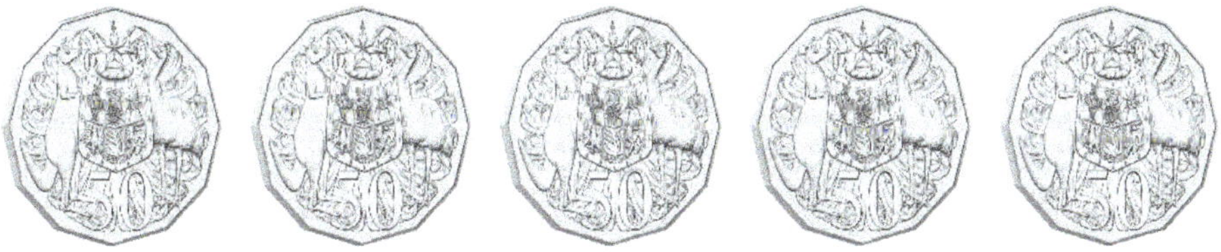

$ _____

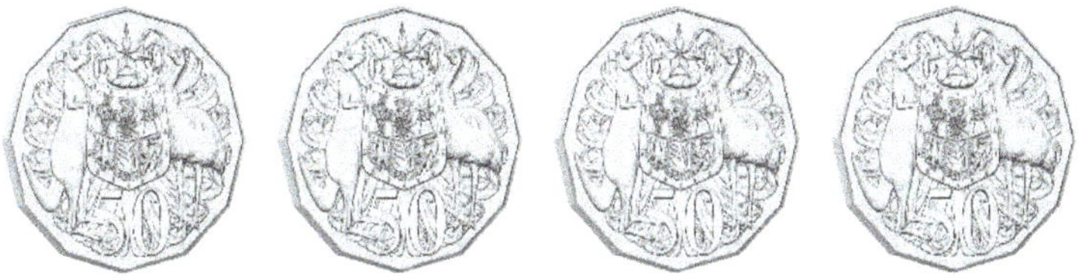

$ _____

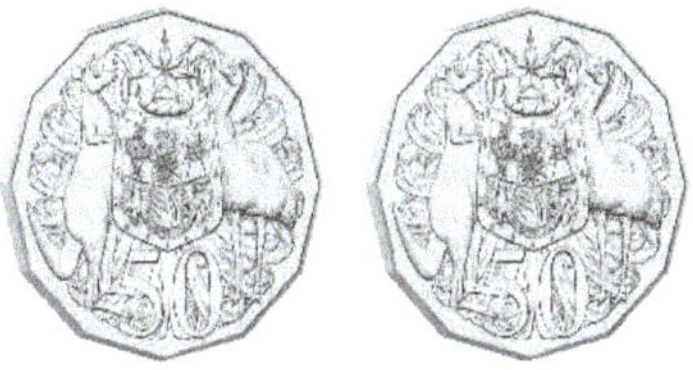

$ _____

Adding $1 and 50c coins

Add each row of coins.

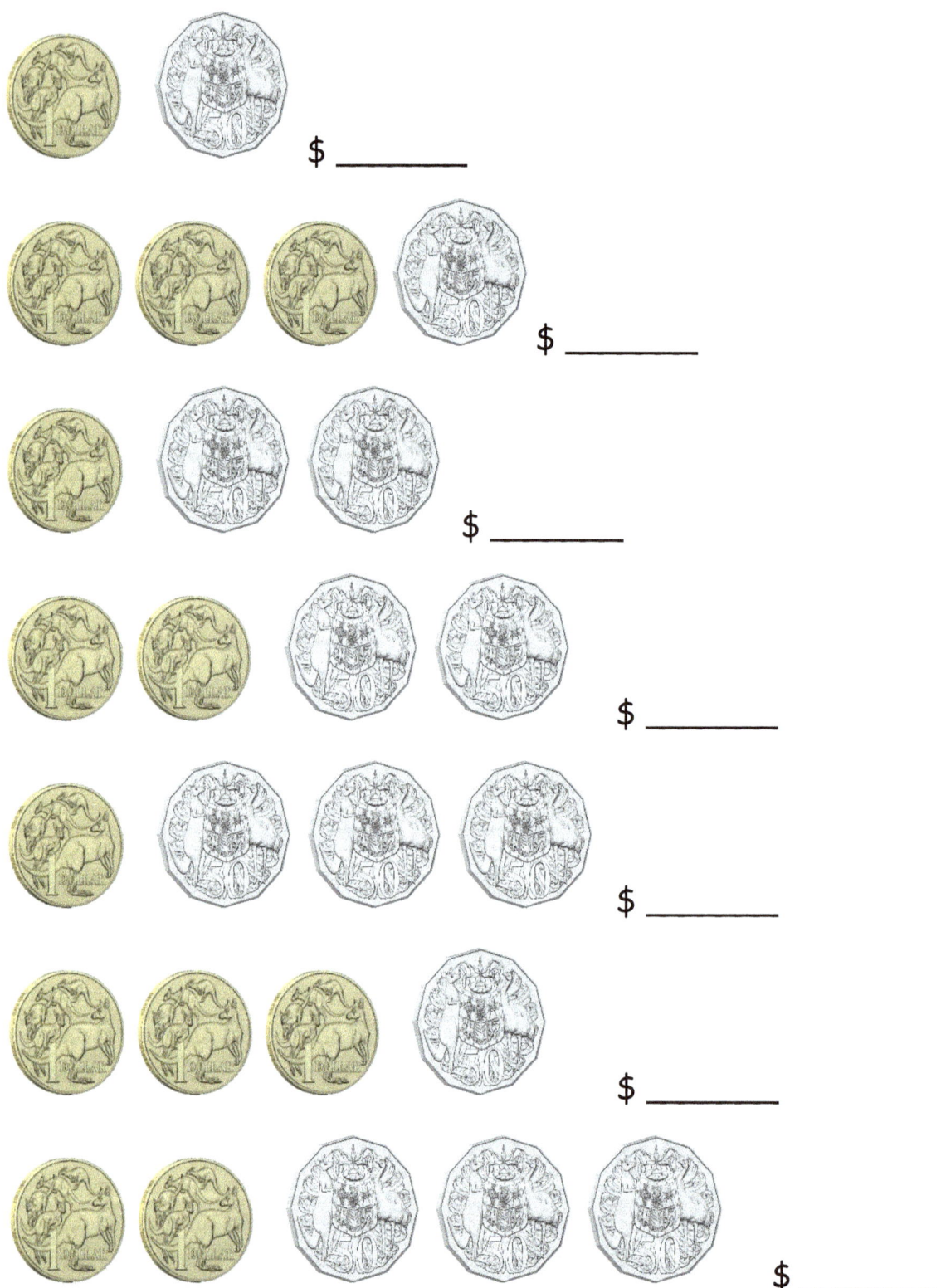

Adding $2, $1 and 50c coins

Add each row of coins.

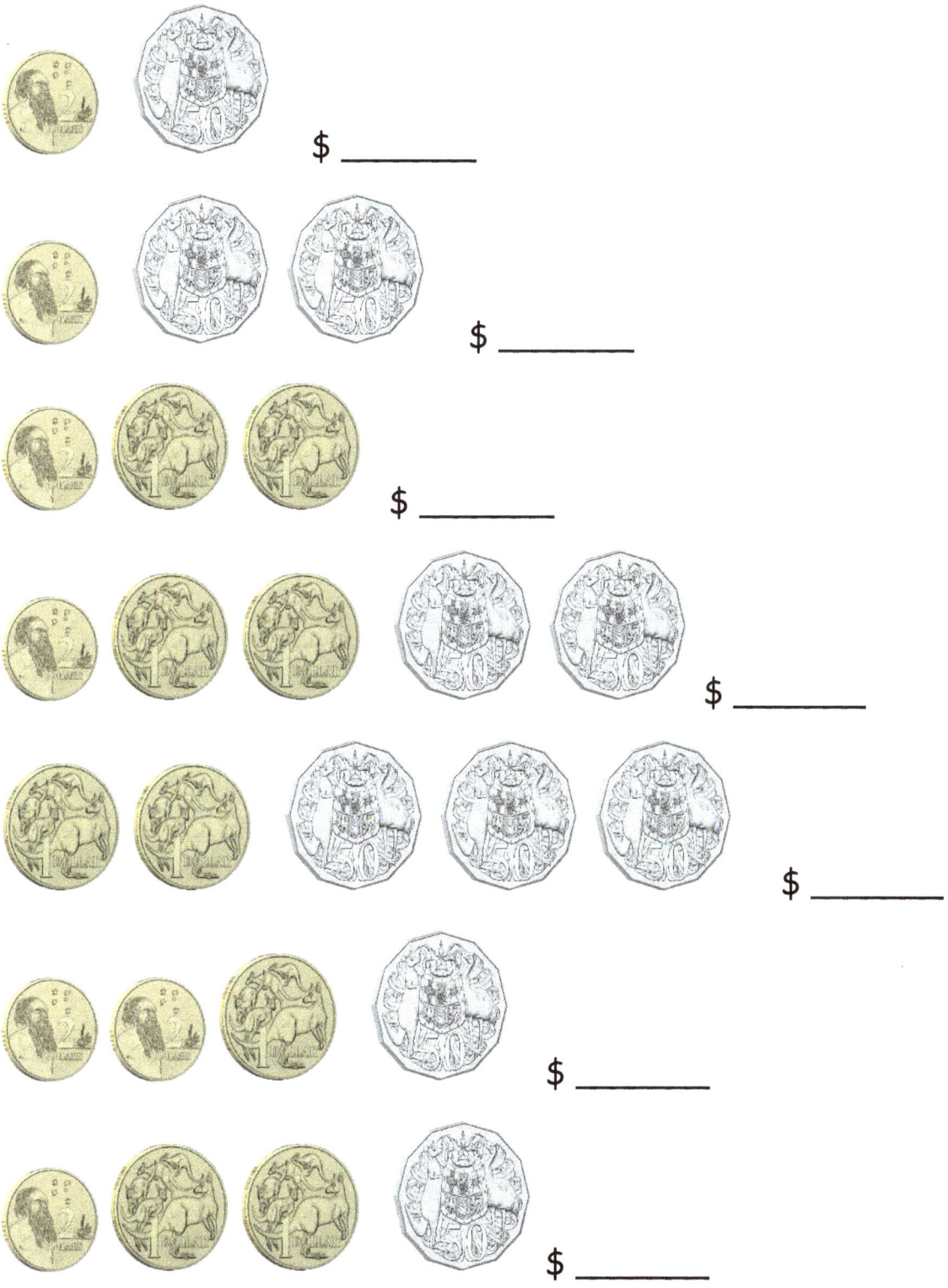

$ _____

$ _____

$ _____

$ _____

$ _____

$ _____

$ _____

Adding $2, $1 and 50c coins

Add each row of coins.

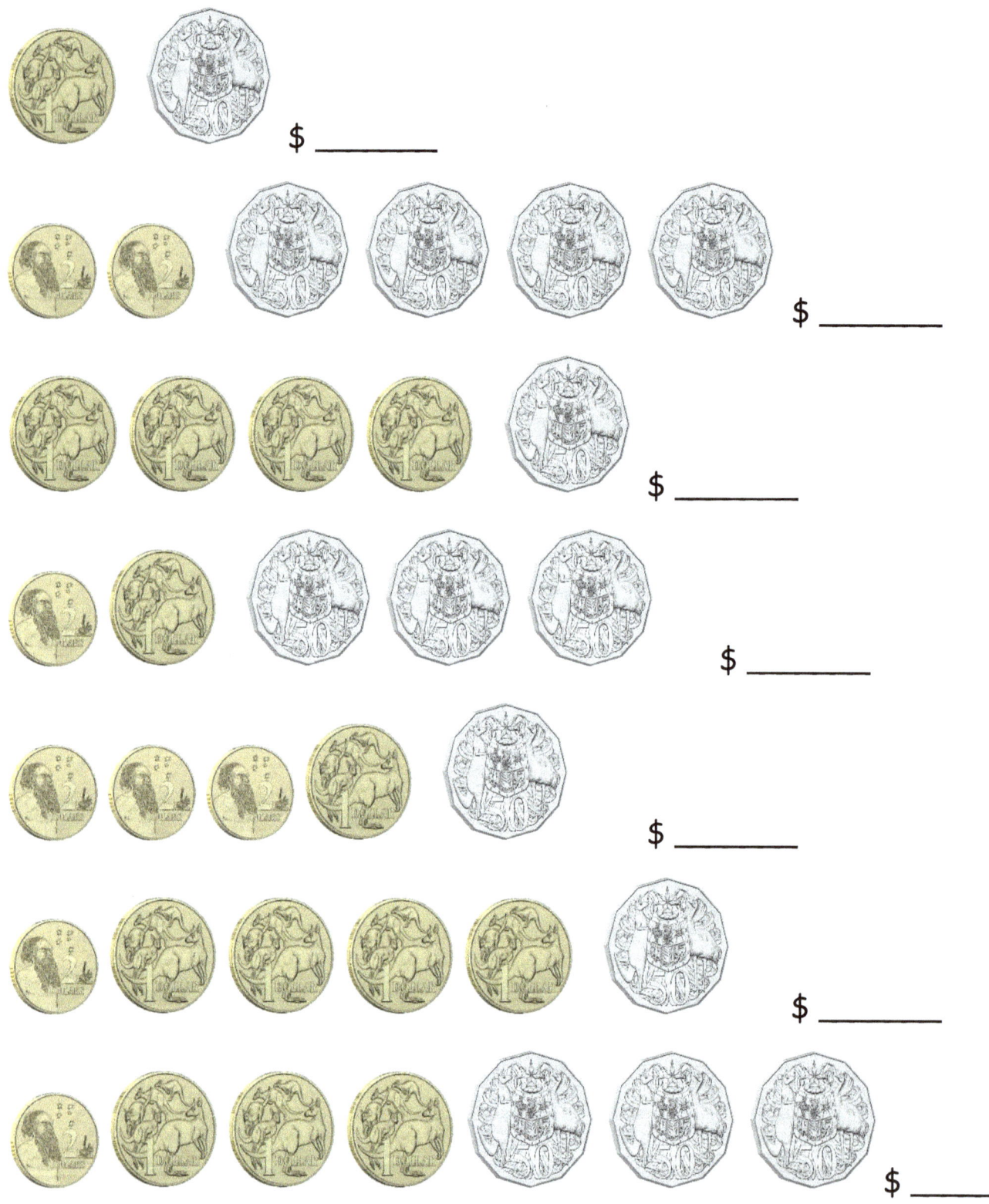

Adding 10c coins

Add each row of 10c coins.

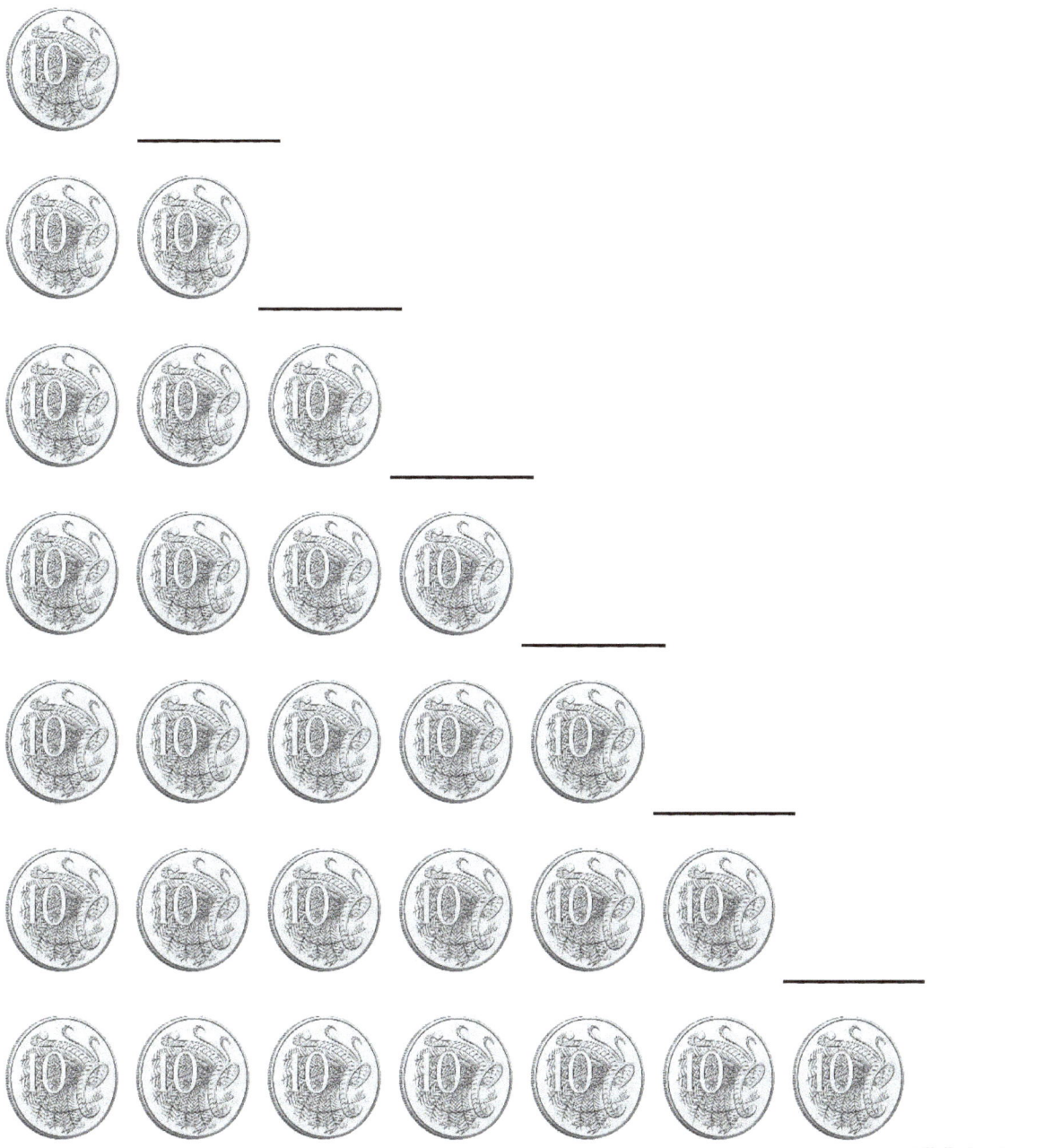

Adding 10c coins

Add each row of 10c coins.

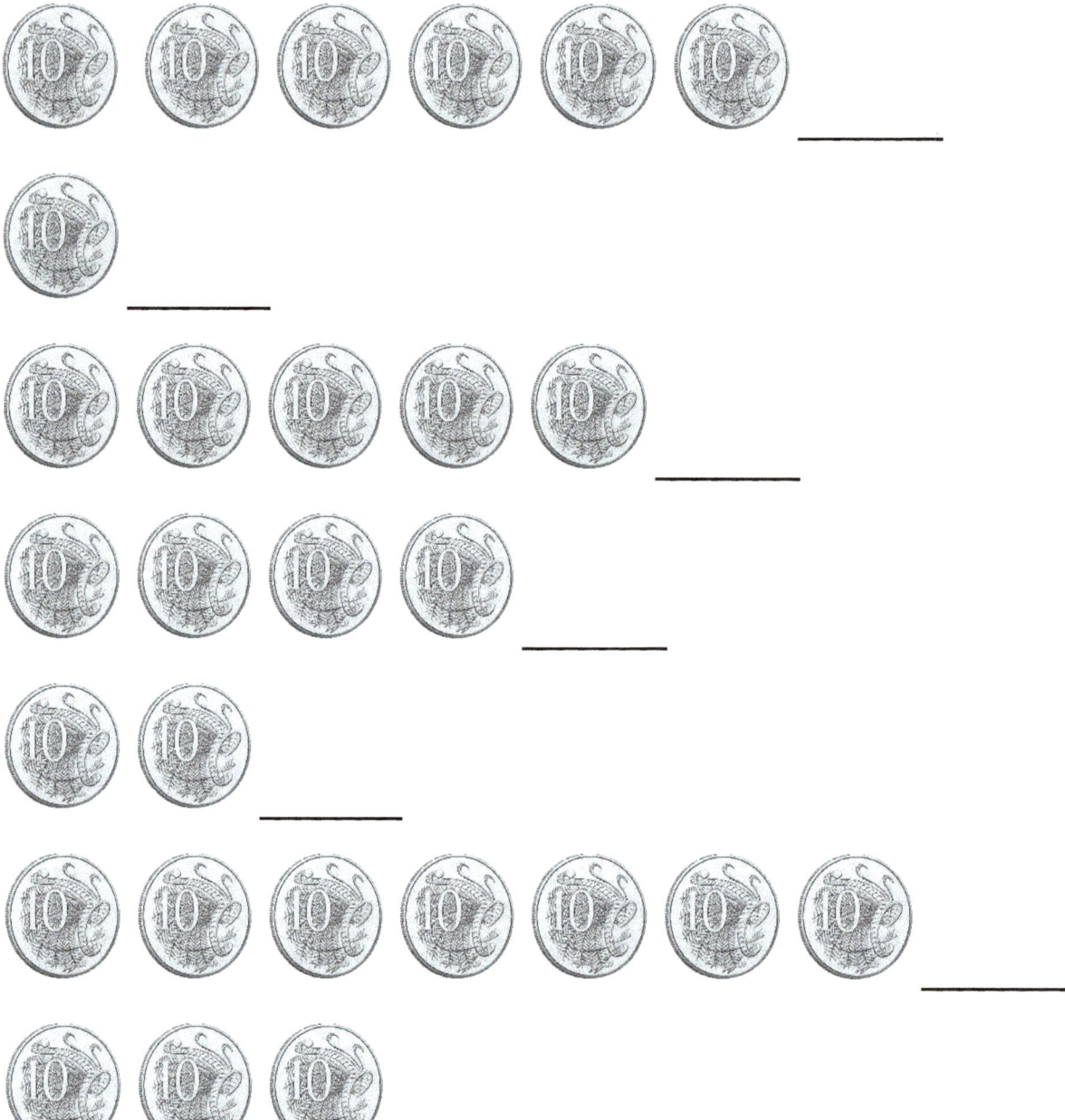

Adding 5c coins

Two five cent coins is the same as one ten cent coin.

Circle groups of two five cent coins and then work out what each row is worth.

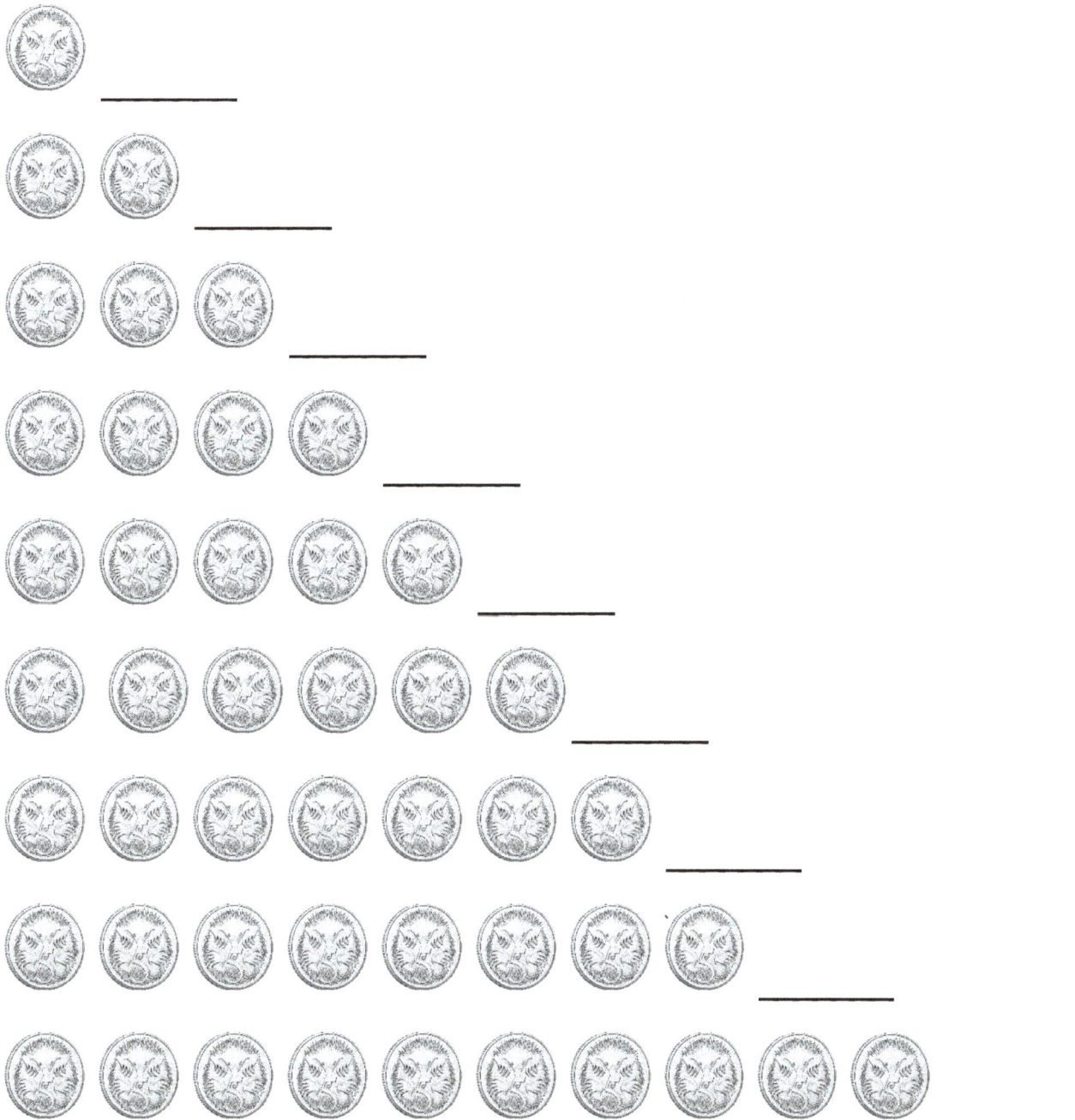

Adding 5c coins

Work out the value of each row of coins.

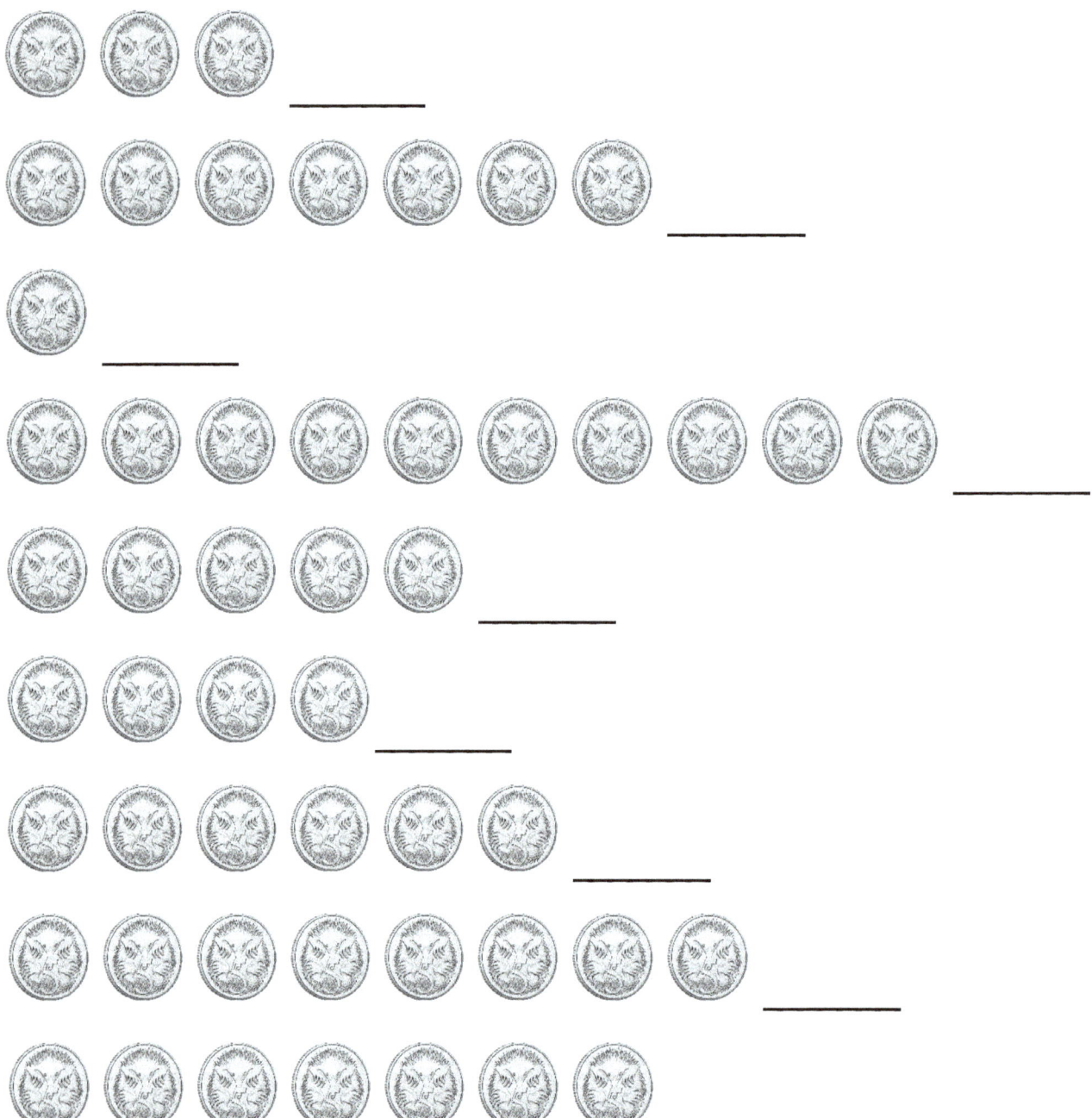

Adding 5c and 10c coins

Circle any groups of 5c coins and then work out the value of each row.

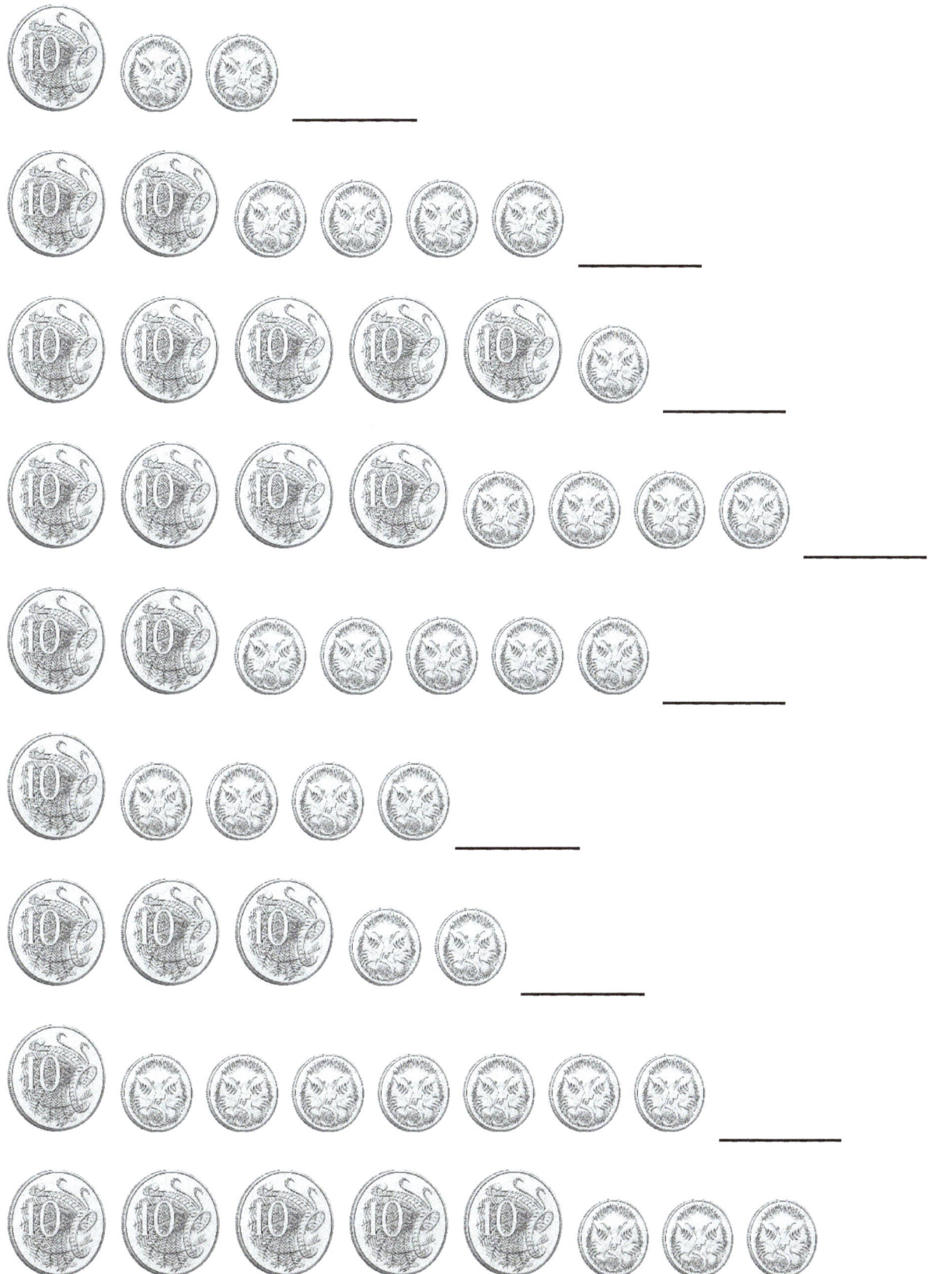

Adding 20c coins

Add each row of 20c coins.

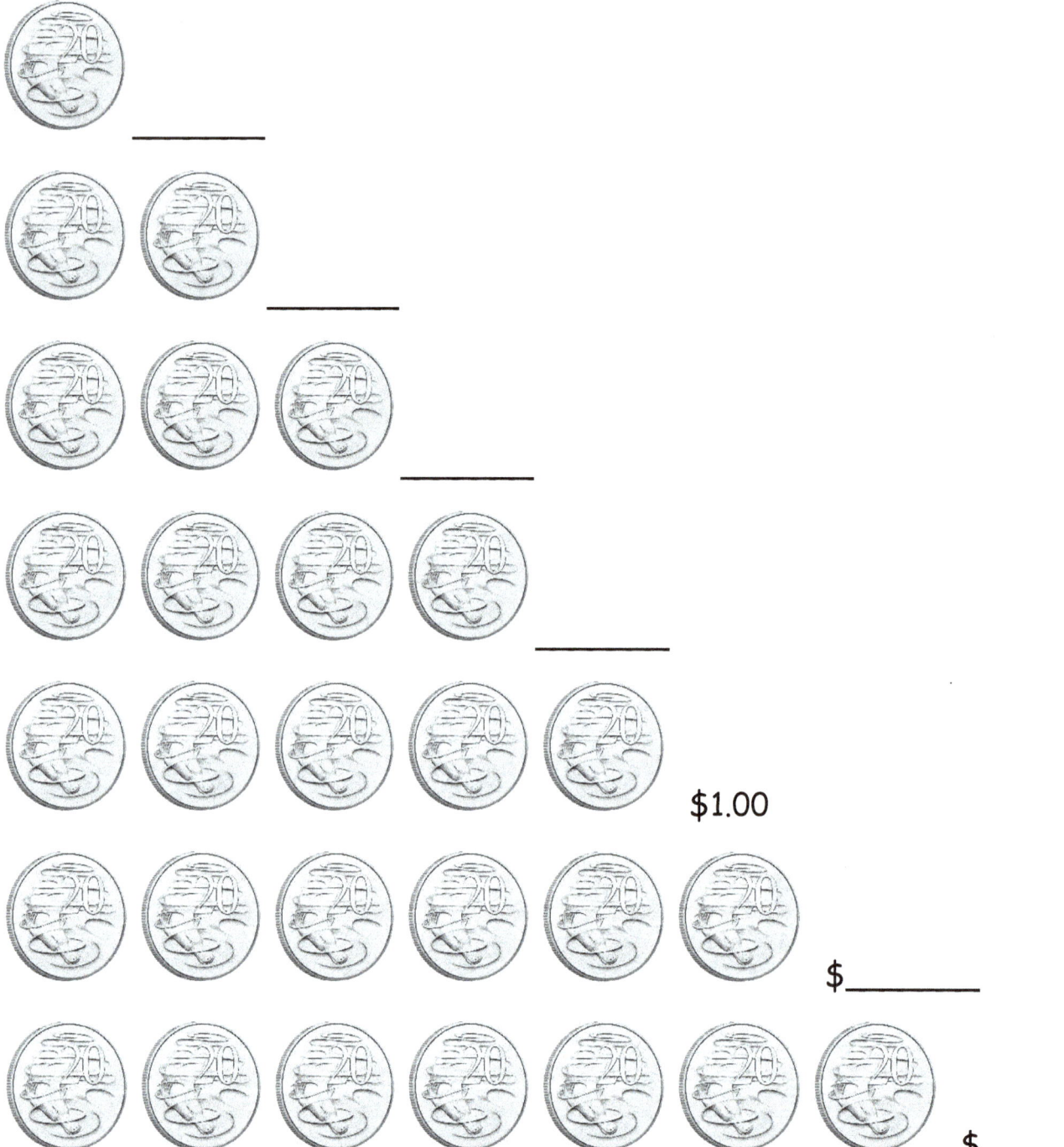

Adding 20c coins

Add each row of 20c coins.

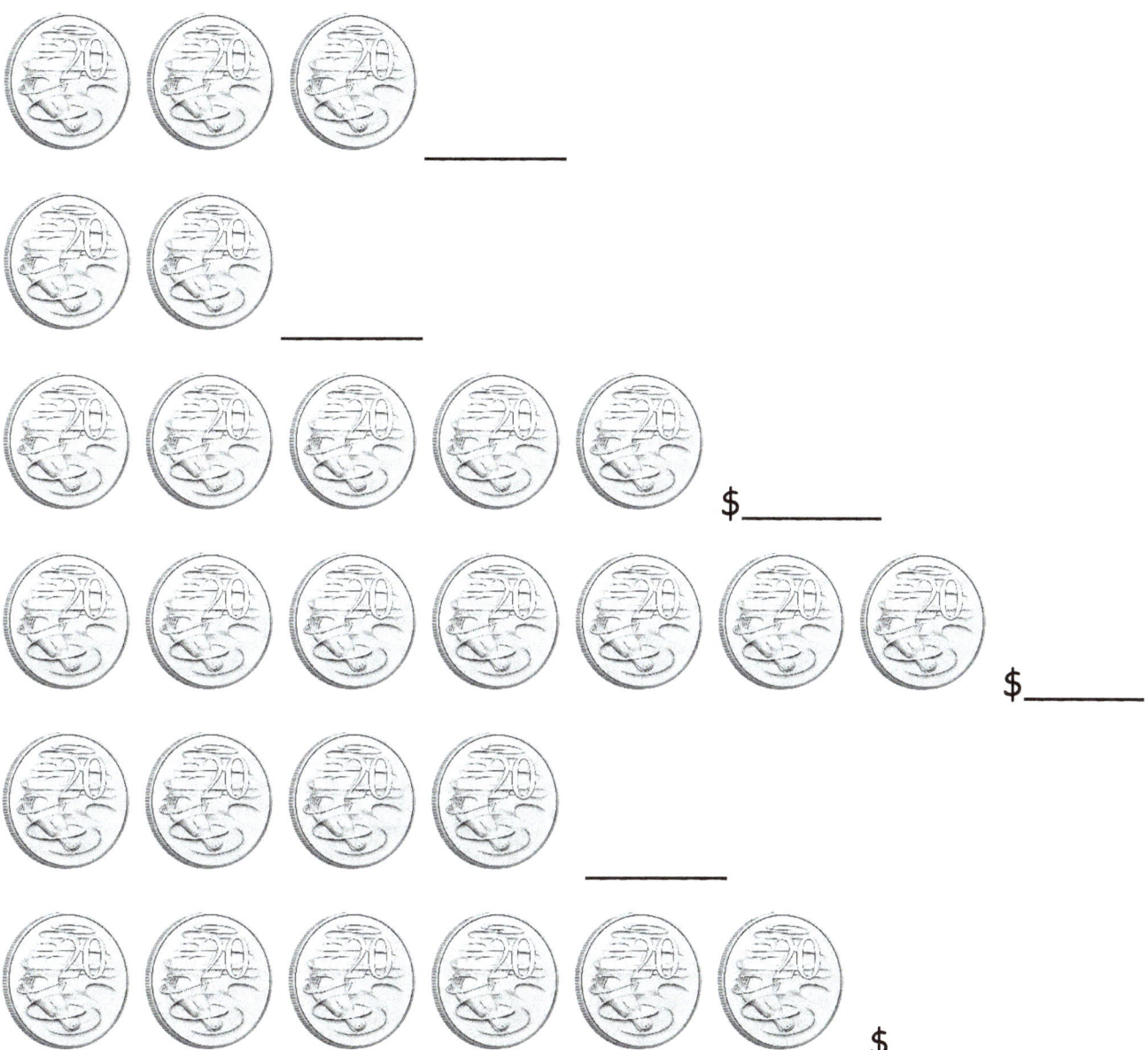

Adding more 20c coins

Five 20c coins equals one dollar. Circle groups of five 20 cent coins then work out how much each set of coins is worth.

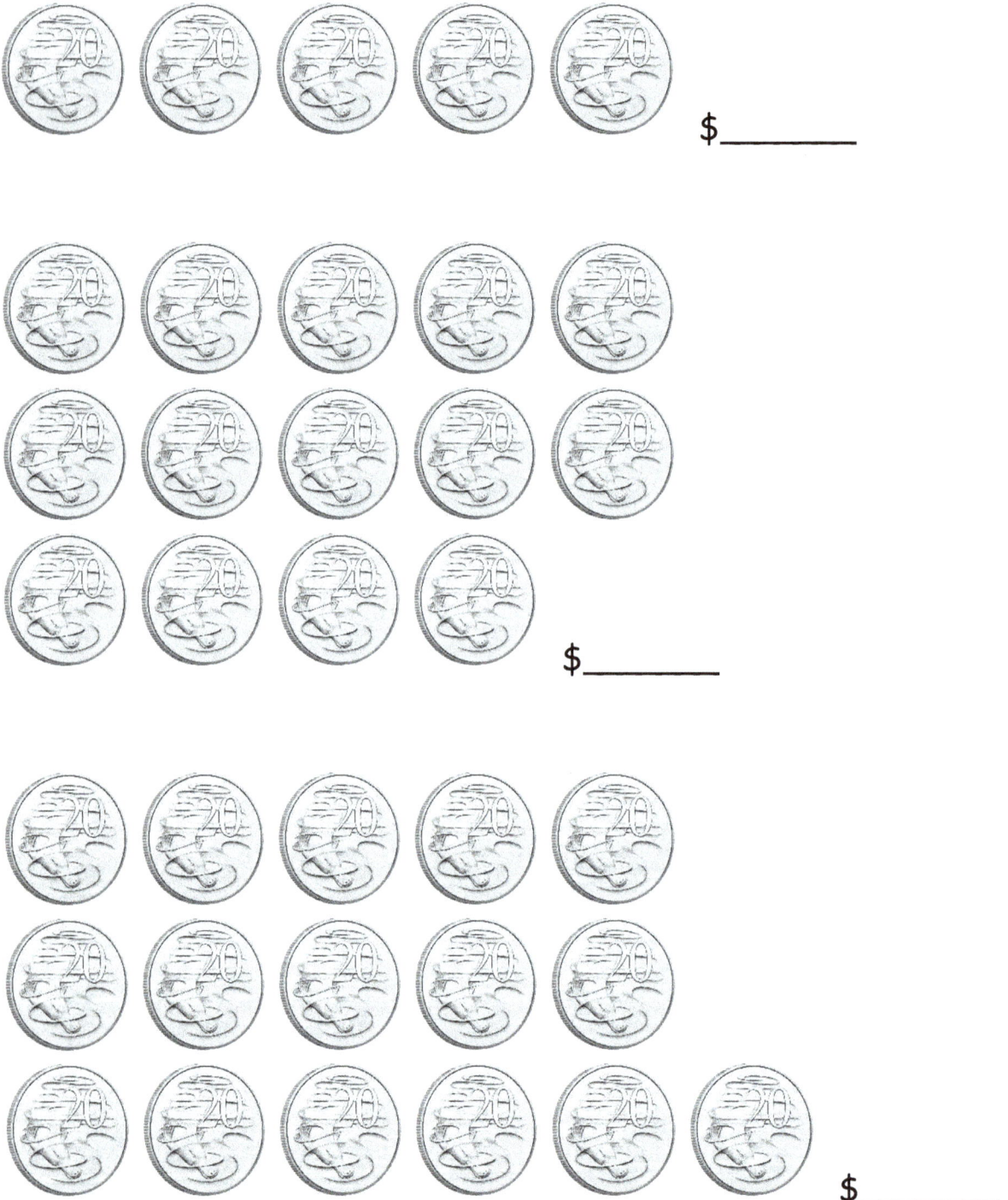

Adding 20c and 10c coins

Add each row of coins.

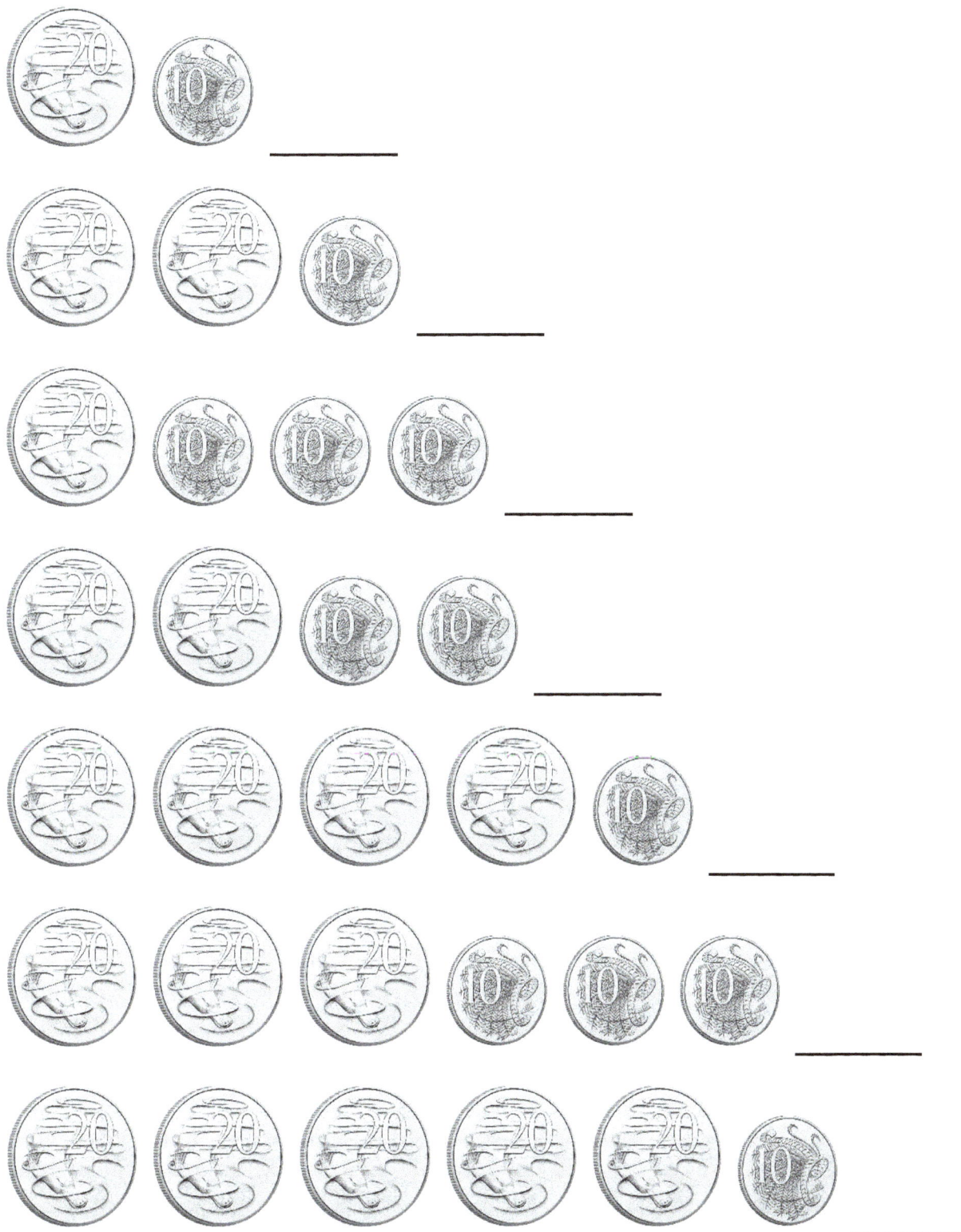

Adding 20c, 10c and 5c coins

Add each row of coins.

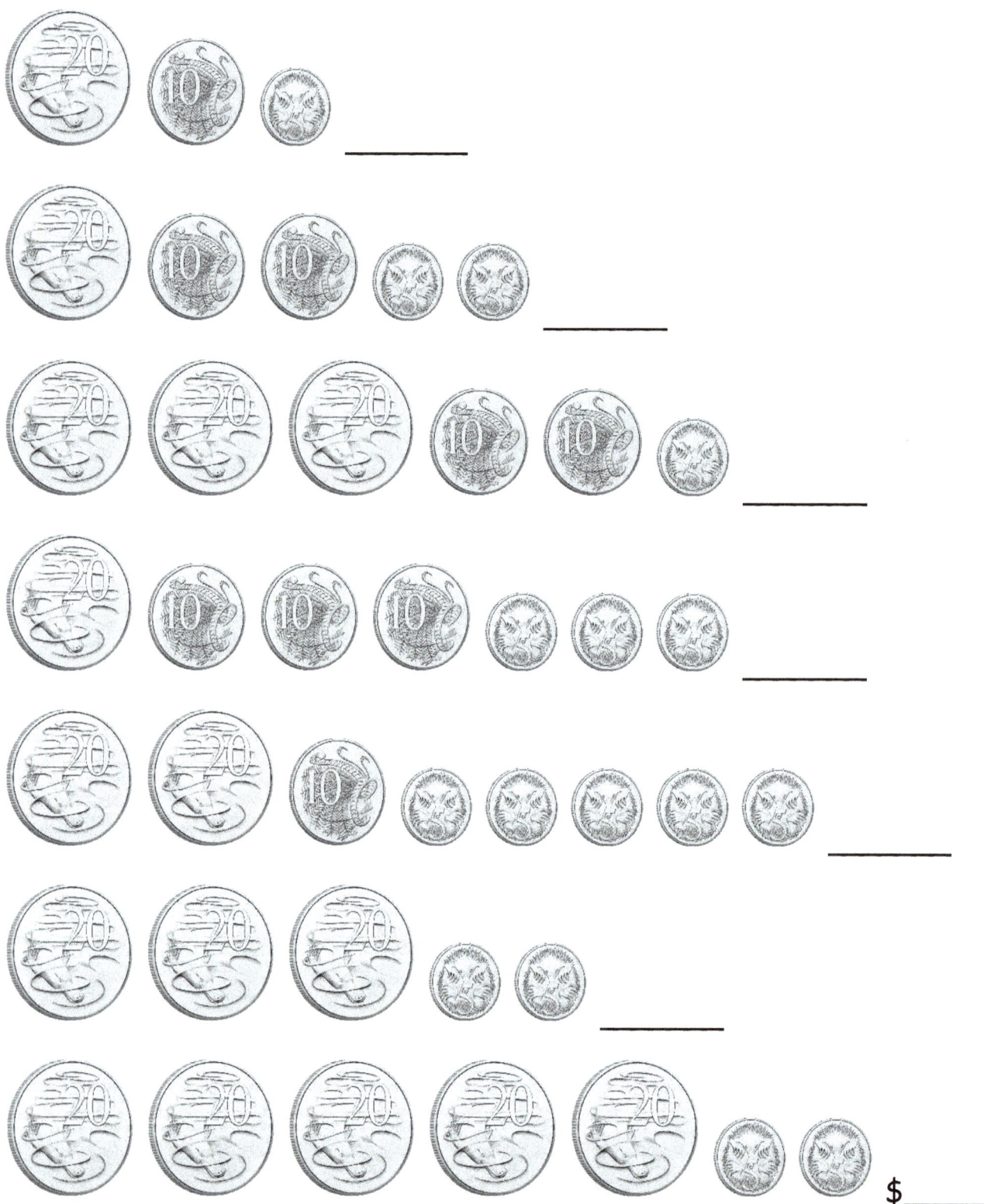

How much in the piggy bank?

Add up the coins in each piggy bank. One great way to do this is to take some pretend coins that match those in the piggy bank, put them in order and then count them from largest to smallest. Count the 5c and 50c coins in groups of two. Don't have any pretend coins? Use the pictures on page 64.

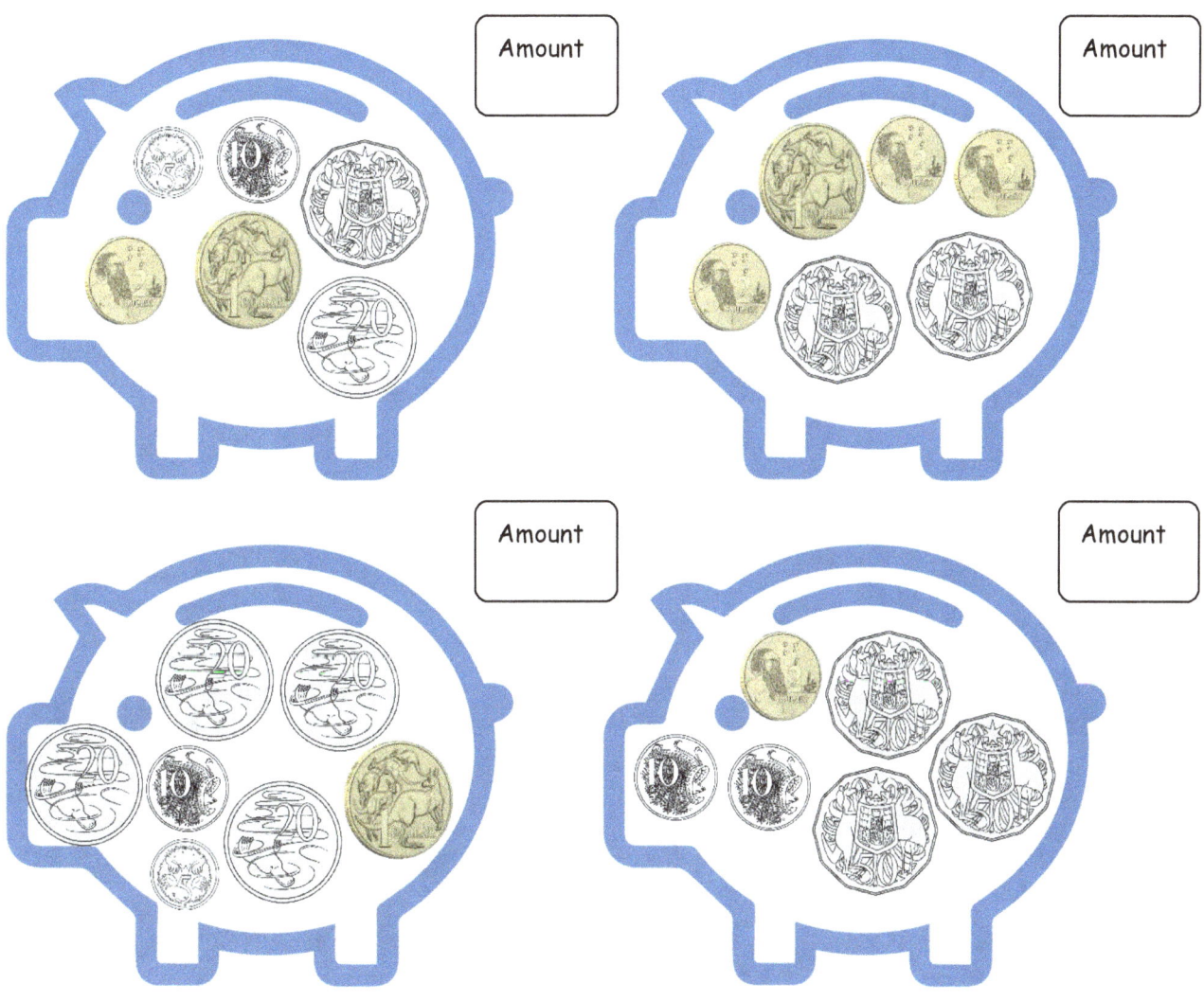

How much money in the piggy banks?

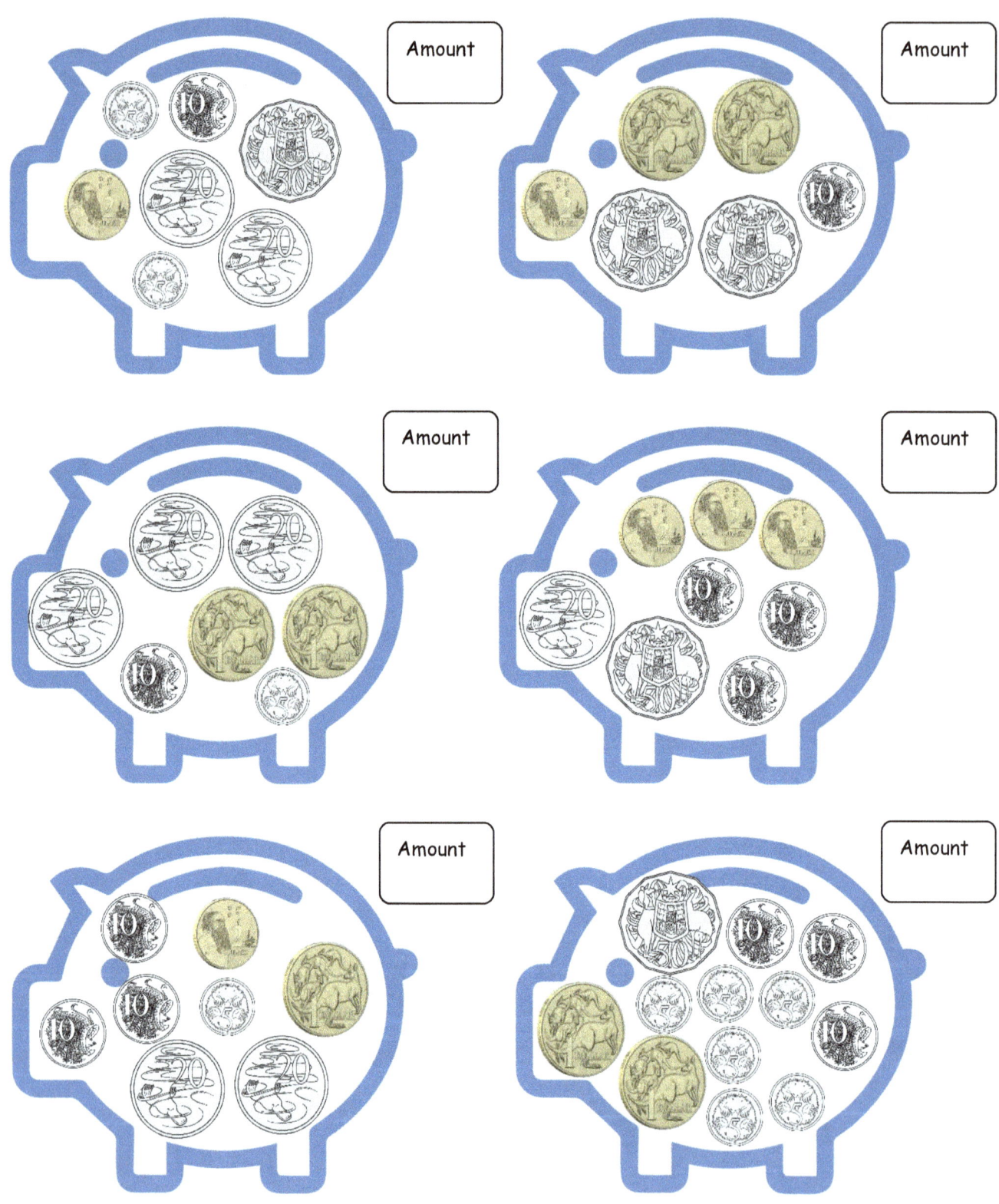

How much money in the money bags?

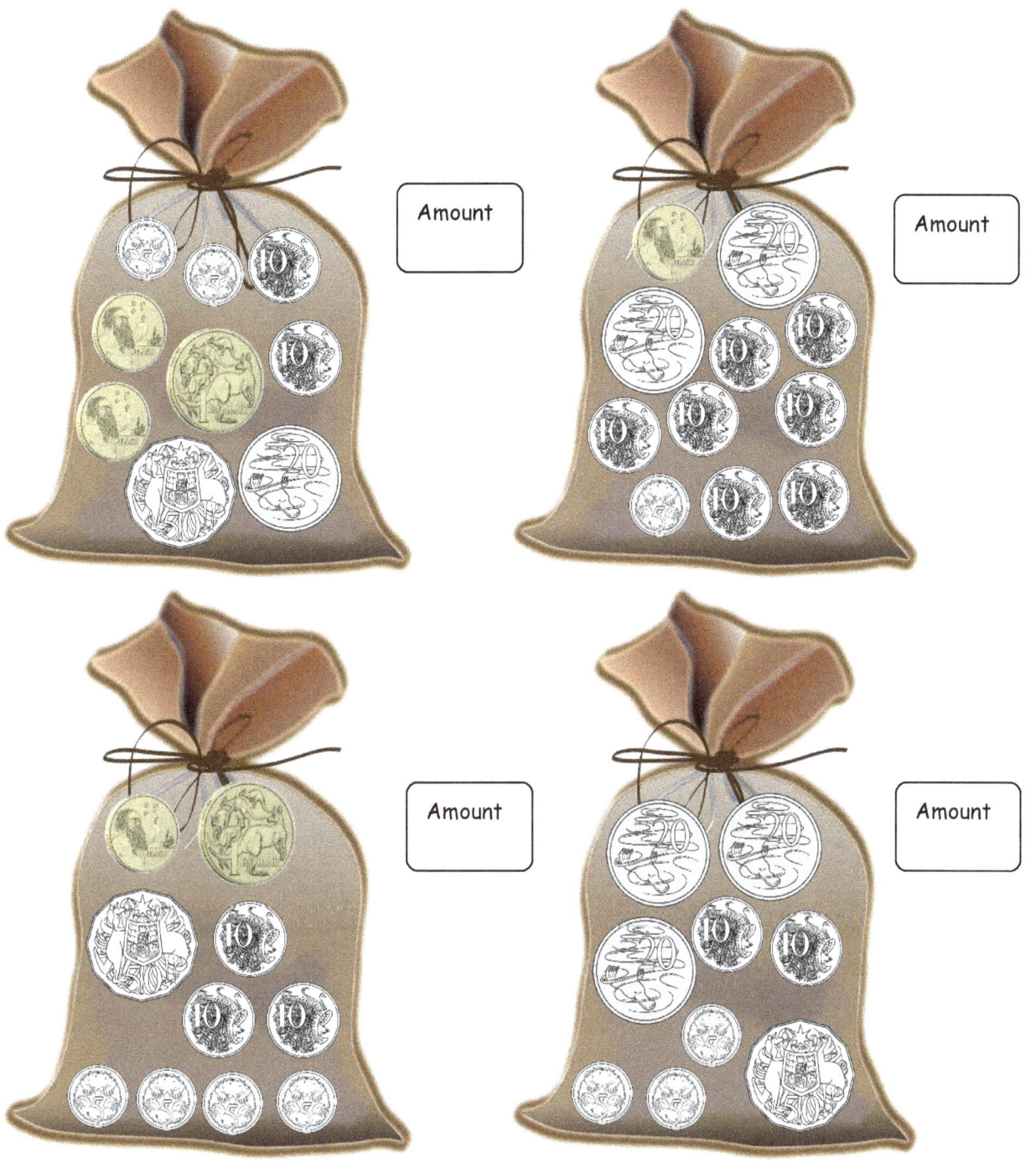

How much money in the money bags?

Making One Dollar

When people count a lot of coins, they will often:

- Make piles of one dollar
- Add up the dollars
- Count the remaining coins and add to the total.

In each of the rows below circle the coins you would put in a pile to make $1. There may be more than one group.

Making One Dollar

In each of the rows below circle the coins you would put in a pile to make $1. There may be more than one group.

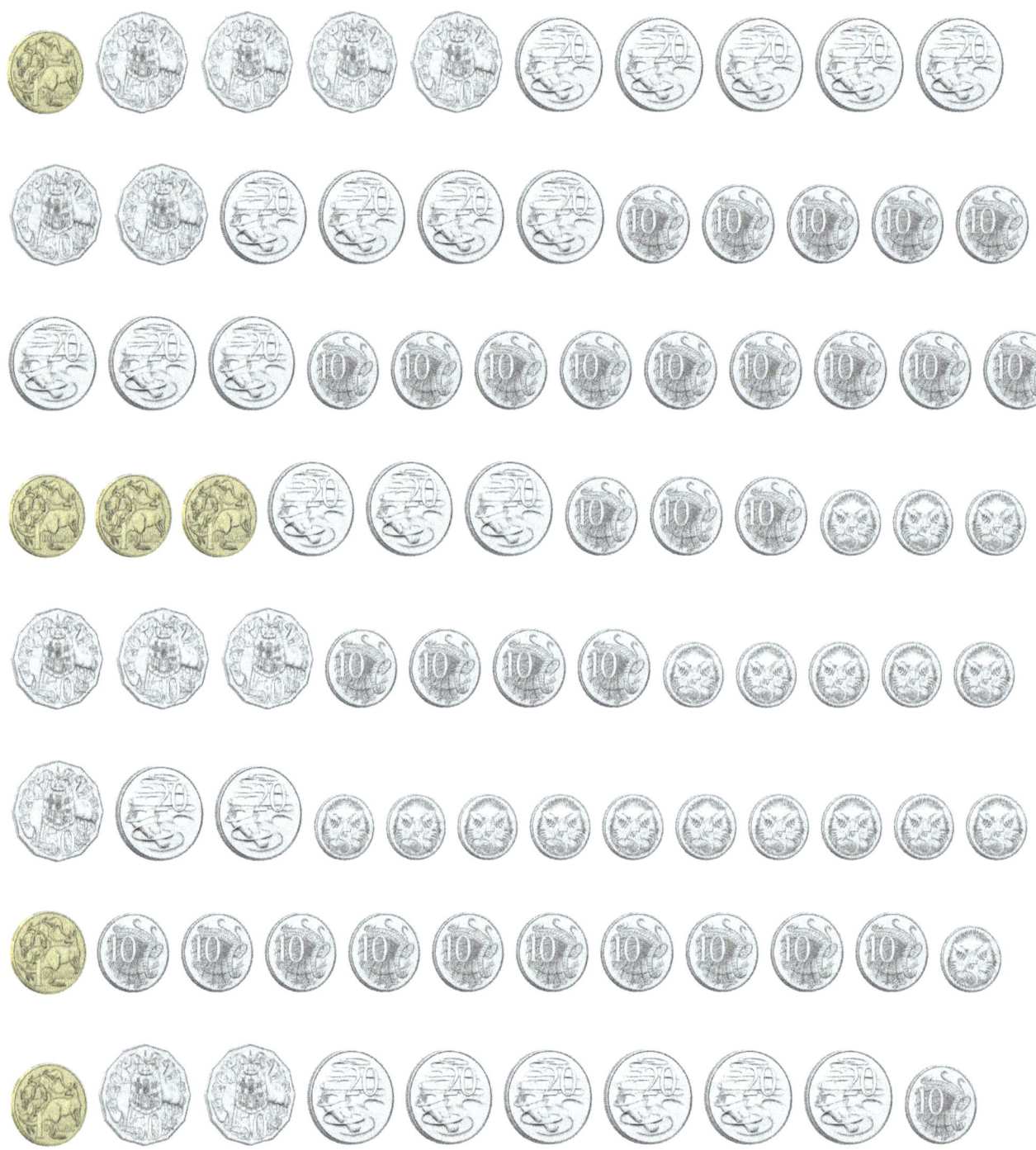

Stocking the shelves

Cut out the pictures on page 62 (or download them from https://warrupress.com/using-australian-money/) Then paste them on the correct shelf below.

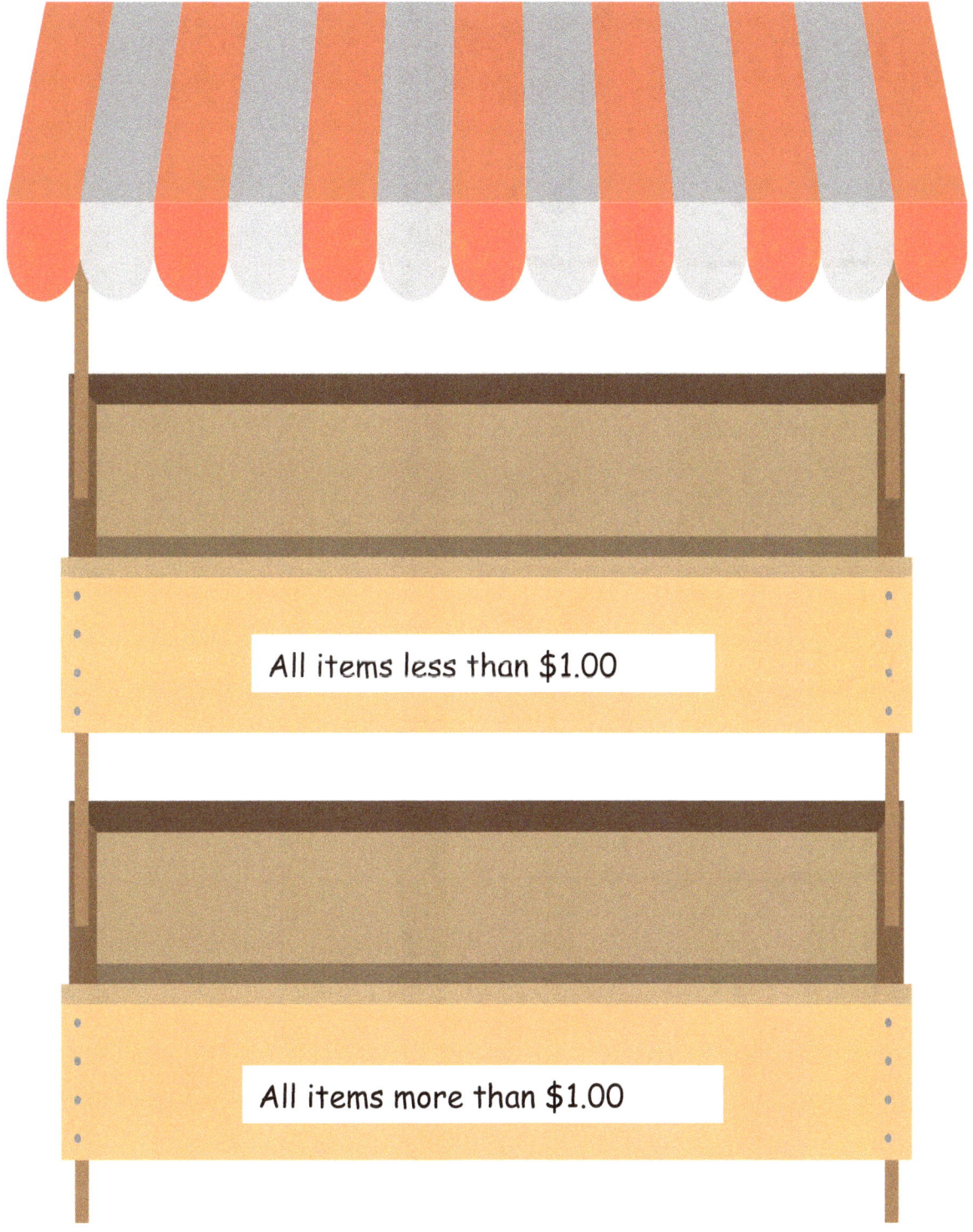

Whole Dollar Amounts

Glue in notes and coins equal to the value of each item from page 62 or
https://warrupress.com/using-australian-money/

A Fast Food Cafe

You have $8. You need to buy food and a drink. What do you order?

Simon buys a hotdog and a bubble tea. How much does it cost him? _____

Freja buys the loaded fries and a milk shake, how much does it cost? _____

Raymond buys a milkshake and two doughnuts. How much does it cost? _____

Using the exact change

Which coins we use to pay for something, depends on what coins we have at the time. There are many ways to make the same amount.

Circle the coins you would use to make the amount on the left.

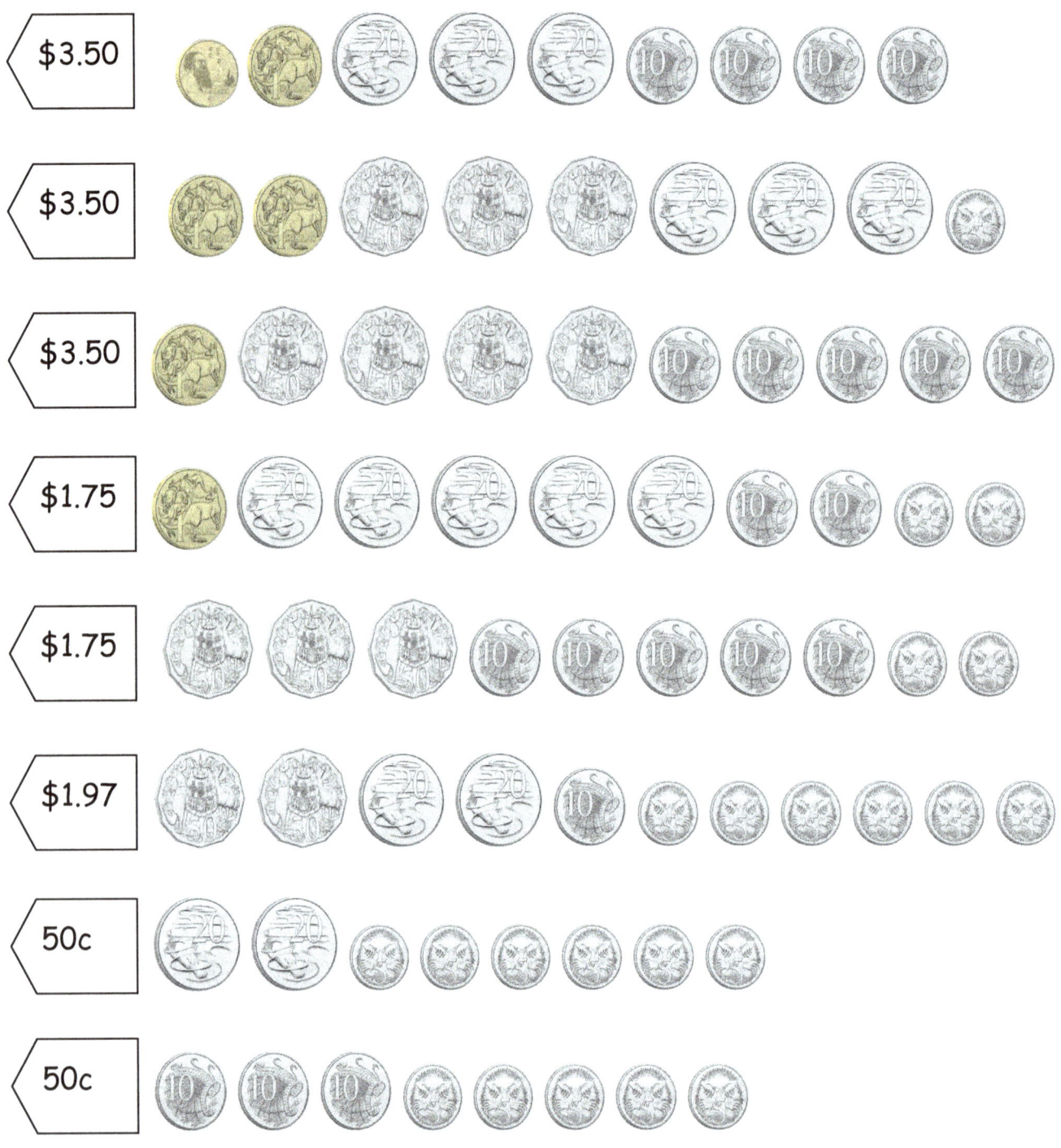

Paying with one coin.

For each of the prices below, circle the smallest coin that is larger or the same as the cost.

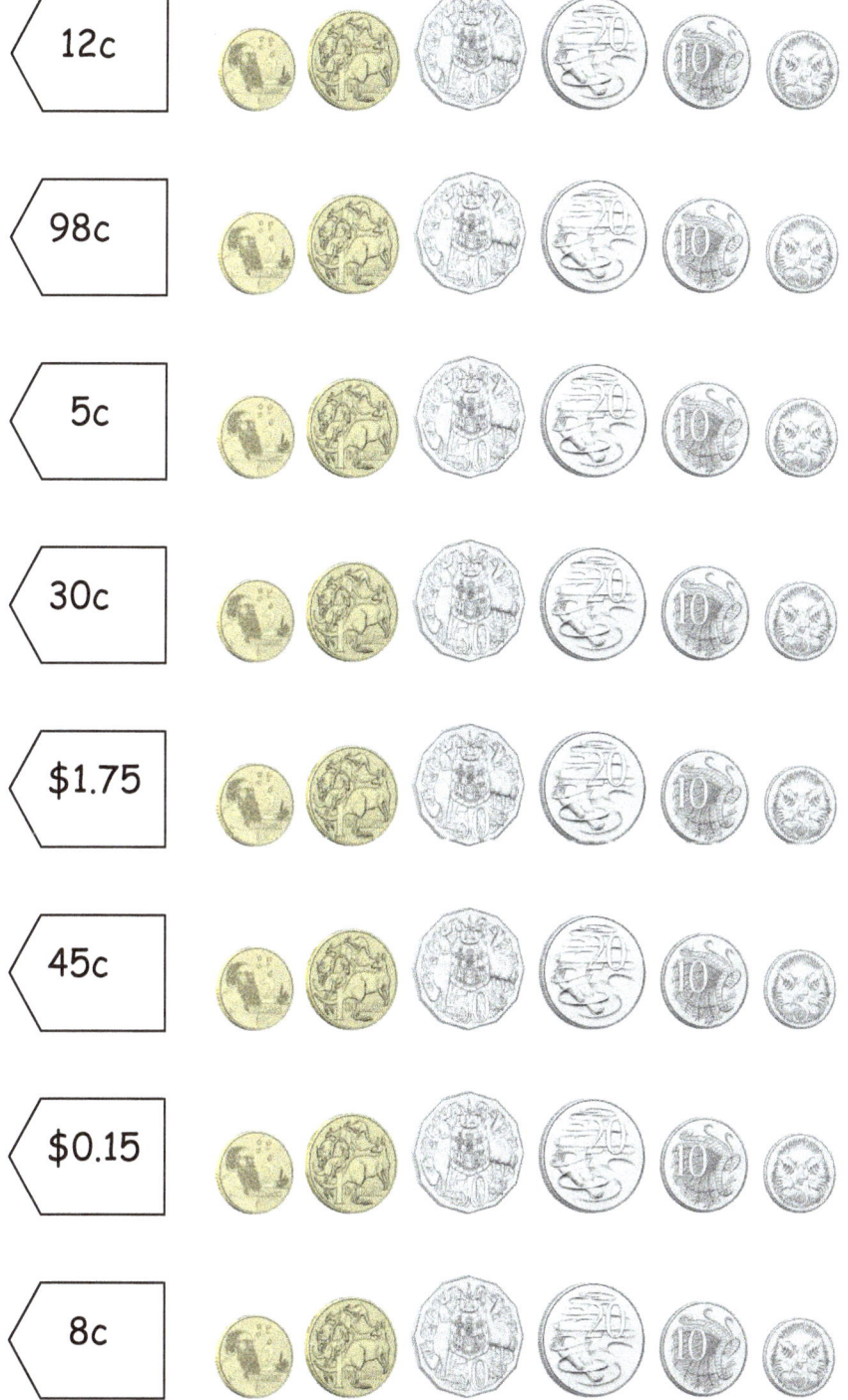

Using the exact change

Circle the coins you would use to make the amount on the left.

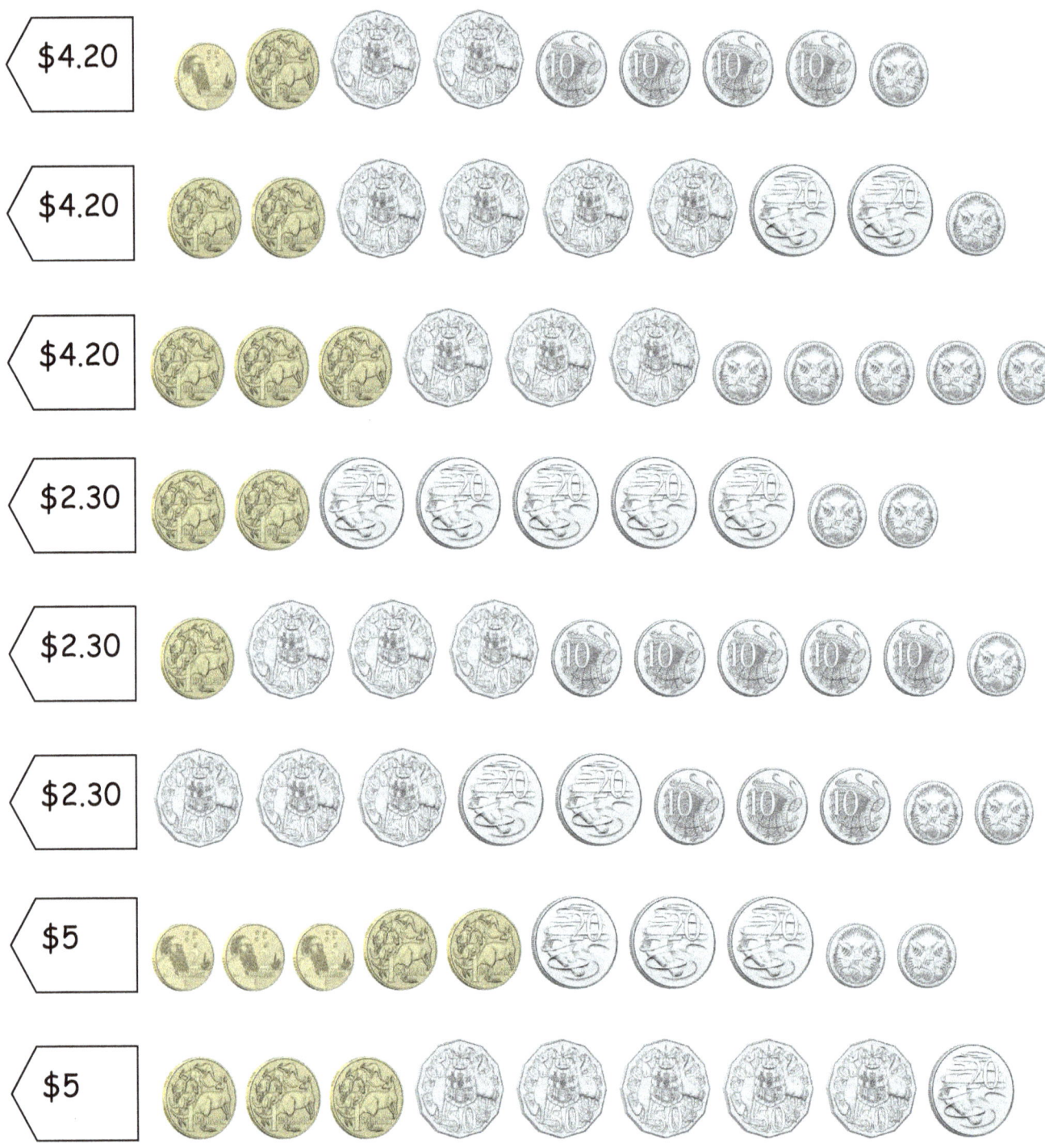

Paying with one or two dollars

If an amount is less than one dollar, you can pay for it with a one dollar coin. Circle the items in green you can buy with a one dollar coin. Circle any remaining items that you could buy with a two dollar coin in purple.

One more dollar

Circle the dollar part of each amount. Write the number that is one more in the box.

One more dollar

Write the number one more than the dollar part of each price. Circle the money that you would need to use to buy the item. The first one is done for you.

Item	One more dollar	Money to use
$35.20	$36	
$21.70		
$5.40		
$8.20		
$3.15		

Which note to use?

For each of the items below, circle the smallest note that is larger than the cost.

Making Ten Dollars

Circle what you need to make ten dollars, in each row.

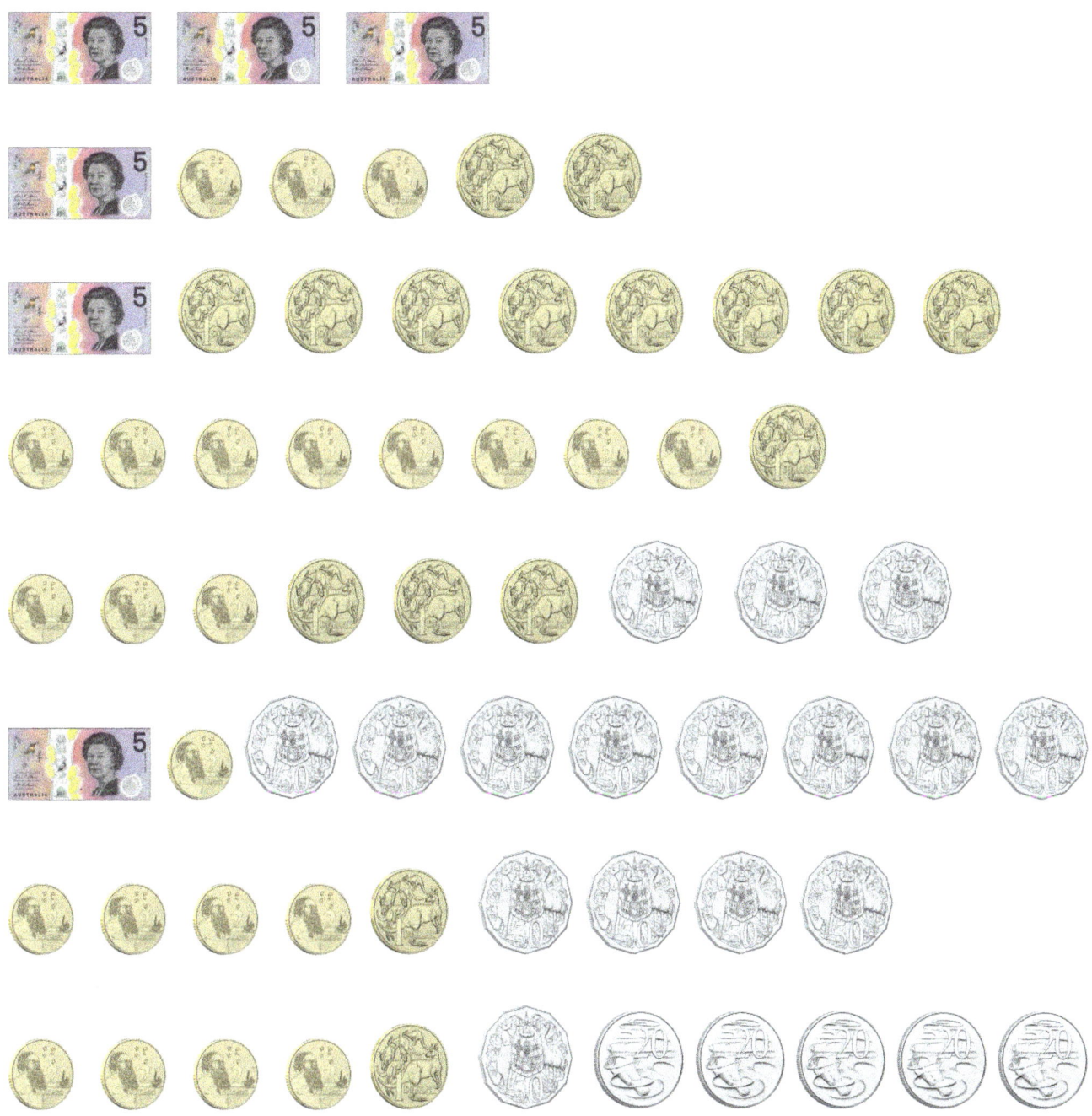

One more dollar

Write the number one more than the dollar part of each price. Circle the money that you would need to use to buy the item.

Item	One more dollar	Money to use
Sweater $69.97		$50, $20, $20, $10, $5, $1, $2
Belt $16.90		$20, $20, $10, $5, $1, $2
Watch $69.50		$20, $20, $20, $10, $10, $5
Sock $2.47		$20, $20, $10, $5, $1, $2
Sandal $22.95		$20, $20, $10, $5, $1, $2
Shoe $137.80		$50, $50, $20, $20, $10, $5, $1, $2

Which notes should I use?

You don't have the correct change, so you need to give more than the price. Circle the note or notes you would give to the checkout operator.

Item	Amount given	Money used
Headphones $36.30		$50, $50, $20, $20, $10, $5
Laptop $97.95		$50, $50, $20, $20, $10, $5
Drink $23		$50, $50, $20, $20, $10, $5
Boot $42.47		$50, $50, $20, $20, $10, $5
Microphone $18		$50, $50, $20, $20, $10, $5
Necklace $12		$50, $50, $20, $20, $10, $5

SPECIAL OFFER
Business Lunch

Pizza or Pasta + **Choose Salad** + **Dessert & Drink**

Margarita — Classic Mix — Tiramisu 250 g *or* Apple Pie 250 g / Coffee 300 ml *or* Ice Tea 300 ml

Italiano — Green Plate — save 50% **$25**

PIZZA

Item	Size	Price
Margarita	450 g	$15
Grand Italia	450 g	$18
Italiano	450 g	$15

PASTA

Italiano	450 g	$15
Ocean Dreams	450 g	$18
Four Seasons	150 g	$15

SALADS

Classic Mix	450 g	$15
Green Plate	450 g	$18
New Cesar	450 g	$15

DRINKS & DESSERTS

Drinks

Soda Water	150 ml	$ 5
Green Tea	250 ml	$ 2
Cappuccino	100 ml	$ 6
Espresso	50 ml	$ 3
Americano	250 ml	$ 7
Tea Pot	600 ml	$ 8
Milk	100 ml	$ 2
Chocolate	150 ml	$ 5
Latte	300 ml	$ 4
Fresh Juice	150 ml	$ 2

Desserts

Item	Size	Price
Cookies	150 g	$ 15
Ice-cream	250 g	$ 10
Apple Pie	100 g	$ 8
Cacao Pie	50 g	$ 12
Tiramisu	250 g	$ 11
Pudding	600 g	$ 14
Cacao Cake	100 g	$ 10
Berry Cake	150 g	$ 8
Chokolate	300 g	$ 15
Donates	150 g	$ 13

Chocolate & Bar

Free home delivery!

Ordering from café

For the business lunch you can choose either the tiramisu or the apple pie. Which one is worth more? _____

It is not lunchtime, so you cannot get the Business Lunch. You have $25. What do you order? _____
_____

How much does your order come to? _____

What are the two cheapest drinks? _____

What is the most expensive dessert? _____

Sonya orders a Grand Italia Pizza and a Green Tea. How much does her order come to? _____

If Sonya has $25, can she order dessert? _____

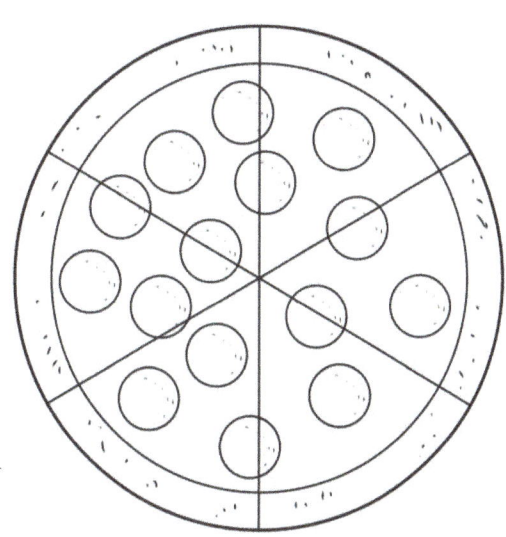

Resources for Exercises

Ordering Australian Money, page 7

Ordering Australian Money, page 9

Images for Stocking the shelves, page 45.

Whole Dollar Amounts on page 46

Notes

This page may be photocopied by the original purchaser. If the size is changed, then the notes must be less than three-quarters or greater than one and a half times the length and width of the genuine banknote.

ISBN: 978-1-922819-04-8 © Warru Press, 2023

Coins

This page may be photocopied by the original purchaser. This is for the sole use of completing the exercises in this book. The purchaser must comply with the criteria set out by the Royal Australian Mint. This page can also be downloaded as a pdf from: https://warrupress.com/using-australian-money/
The pdf must not be altered.

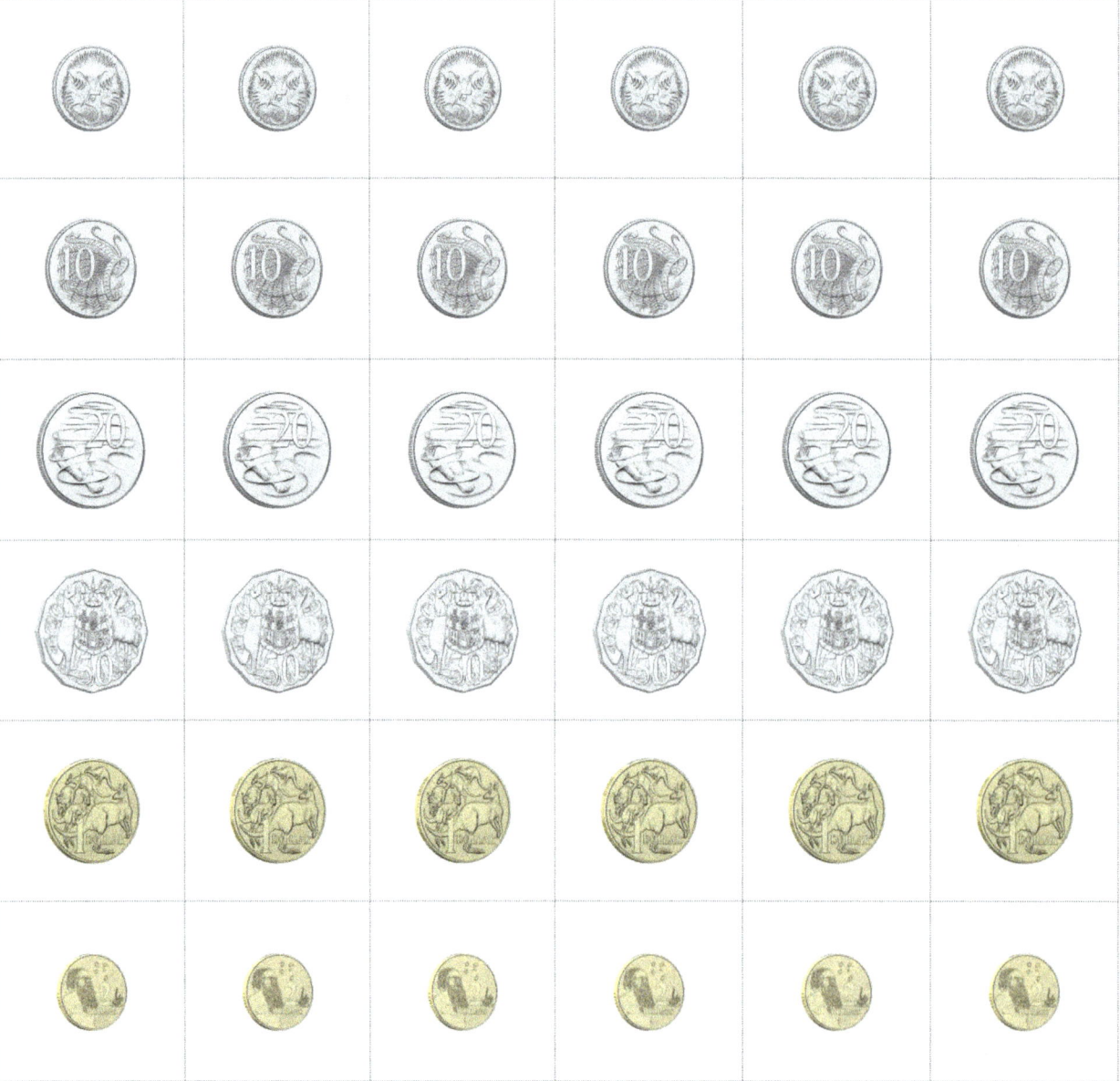

Answers

Page 3
Teacher or parent check

Page 4
Gold coins: 15
Silver coins: 21

Page 5
5c: 7
10c: 7
20c: 7
50c: 5
$1: 9
$2: 6

Page 7 to 9
Teacher or parent check

Page 10
$1, $2, $3, $4, $5, $6

Page 11
$4, $1, $5, $3, $2, $8

Page 12
$2, $4, $6, $8, $10, $12

Page 13
$8, $4, $10, $6, $12, $10, $2

Page 14
$3, $5, $4, $7, 7, $9

Page 15
Teacher or parent check

Page 16
First column: $5, $9, $4, $7, $10, $11, $19, $15, $37,
Second column: $1, $8, $2, $6, $3, $20, $17, $21, $30, $100

Page 17
3, 1, 8, 4, 5, 7, 0, 1, 9, 9

Page 18
$17, $29, $46, $58, $82

Page 19
91c, 39c, 81c, 13c, 68c, 49c, 87c, 26c, 51c, 73c

Page 20
$1.00, $7.00, $8.00, $19.00, $47.00, $56.00, $62.00 $3, $6, $12, $27, $38

Page 21
$\frac{51}{100}$, 0.51, $\frac{13}{100}$, 0.13

Page 22
$\frac{26}{100}$, 0.26, $\frac{48}{100}$, 0.48

$\frac{56}{100}$, 0.56, $\frac{47}{100}$, 0.47

$\frac{25}{100}$, 0.25, $\frac{68}{100}$, 0.68

Page 23
$0.72, $0.99, $0.42, $0.27, $0.13, $0.30, $0.05, 68c, 15c, 50c, 24c, 82c, 22c, 75c, 2c

Page 24
$1, $1.50, $2, $2.50

Page 25
$0.50, $1.50, $2.50, $2, $1

Page 26
$1.50, $3.50, $2, $3, $2.50, $3.50, $3.50

Page 27
$2.50, $3, $4, $5, $3.50, $5.50, $4.50

Page 28
$1.50, $6, $4.50, $4.50, $7.50, $6.50, $6.50

Page 29
10c, 20c, 30c, 40c, 50c, 60c, 70c

Page 30
60c, 10c, 50c, 40c, 20c, 70c, 30c

Page 31
5c, 10c, 15c, 20c, 25c, 30c, 35c, 40c, 50c

Page 32
15c, 35c, 5c, 50c, 25c, 20c, 30c, 40c, 35c

Page 33
20c, 40c, 55c, 60c, 45c, 30c, 40c, 45c, 65c

Page 34
20c, 40c, 60c, 80c, $1.00, $1.20, $1.40

Page 35
60c, 40c, $1.00, $1.40, 80c, $1.20

Page 36
$1.00, $2.80, $3.20

Page 37
30c, 50c, 50c, 60c, 90c, 90c, $1.10

Page 38
35c, 50c, 85c, 65c, 75c, 70c, $1.10

Page 39
1st row: $3.85, $8
2nd row: $1.95, $3.70

Page 40
1st row: $3.10, $5.10
2nd row: $2.75, $7
3rd row: $3.75 $3.10

Page 41
1st row: $3.40, $3
2nd row: $9, $3.15
3rd row: $3.80, 95c

Pages 42 to 46
Teacher or parent check

Page 47
Answers may vary, $10, $10, $8

Page 48
Teacher or parent check

Page 49
20c, $1, 5c, 50c, $2, 50c, 20c, 10c

Pages 50 & 51
Teacher or parent check

Page 52
1st row: $10, $12, $8,
2nd row: $9, $5, $4
3rd row: $9, $6, $2

Page 53
$22, $6, $9, $4

Page 54
$10, $10, $20, $10, $5, $5

Page 55
Teacher or parent check

Page 56
$70, $17, $70, $3, $23, $138

Page 57
$40, $100, $25, $45, $20, $15

Page 58
Tiramisu, answers will vary, green tea and fresh juice, cookies or chocolate, $20, no

Copyright

The images of money in this book licensed by Dancing Crayon Designs.
© www.DancingCrayon.com

The supermarket shelves on page 45 are licensed from 123rf.com.

The image of grocery items on page 62 is designed by macrovector / Freepik

The bakery menu on page 46 is designed by macrovector / Freepik

The Fast Food Menu on page 47 is designed by Freepik.

The items in One More Dollar on pages 56 and some of the items on page 57 are by Kate Hadfield Designs

The Australian Animals are by Becky Beach of PLR Beach.

Other images in this book are licenced from DepositPhotos.

www.ingramcontent.com/pod-product-compliance
Lightning Source LLC
Chambersburg PA
CBHW082009090526
44590CB00020B/3413